# The Emergence of Everything

# The Emergence of Everything

## HOW THE WORLD BECAME COMPLEX

Harold J. Morowitz

OXFORD
UNIVERSITY PRESS
2002

# OXFORD
UNIVERSITY PRESS

Oxford   New York
Auckland   Bangkok   Buenos Aires   Cape Town
Chennai   Dar es Salaam   Delhi   Hong Kong   Istanbul   Karachi   Kolkata
Kuala Lumpur   Madrid   Melbourne   Mexico City   Mumbai   Nairobi
São Paulo   Shanghai   Singapore   Taipei   Tokyo   Toronto

and an associated company in Berlin

Copyright © 2002 by Harold J. Morowitz

Published by Oxford University Press, Inc.
198 Madison Avenue, New York, New York 10016
www.oup.com

Oxford is a registered trademark of Oxford University Press

**Library of Congress Cataloging-in-Publication Data**
Morowitz, Harold J.
The emergence of everything : how the world became complex / Harold J. Morowitz.
p.   cm.
ISBN 0-19-513513-X
1. Evolution.   2. Complexity (Philosophy)   3. Science—Philosophy.   I. Title.
Q175.32.E85 M67 2002
116—dc21          2002070388

1   3   5   7   9   8   6   4   2

Printed in the United States of America
on acid-free paper

# Preface

This book began as an attempt to reify the concept of emergence by finding observed examples and looking for defining features and similarities. The emphasis was on emergences in nature as distinguished from the examples that can be generated almost without limit in computer modeling of complex systems. Rather than selecting cases at random, I chose a set that constituted a temporal array from the beginnings of the known universe to the most human of activities. These were somewhat arbitrarily divided into 28 cases. The intent was a more detailed view of the character of each emergence.

While pondering the cases I had chosen, I continued to peruse the journals *Science* and *Nature*. Almost every week I found at least one paper of significance in exploring one or more emergences. It became clear that the original goal was too ambitious. The detailed analysis of each emergence, while desirable, was far too unrealistic. I decided to settle for a broader view and try to get the "big picture" of emergences. Therefore, I apologize to the experts for such a fleeting view of each example. I am reminded of Herman Melville's description of his system of cetology: "The whole book is but a draft—nay, but the draft of a draft."

We are clearly at the beginning of viewing science from the new perspective of emergence. I believe that it will provide insights into the evolutionary unfolding of our universe, our solar system, our biota, and our humanity. This essay is to introduce some of the concepts that are coming into focus. The outlook is largely scientific, but certain more philosophical and theological elements keep appearing. I offer no apology. I have a deep belief in monism, a world ultimately comprehended by a unified path of understanding. It is the same world on Monday through Thursday as it is on Friday, Saturday, and Sunday.

This book owes a debt to everyone who has shared a dialog with me on the subject matter, to those who have read portions and commented, and to those who have shared the quest. I list the following, with some trepidation that other names have been momentarily overlooked: Ann Butler, James Trefil, Ann Palkovich, James Olds, Robert Hazen, Rob Shumaker, Barbara Given, Lev Vekker, Neil Manson, Karl Stephan, James Barham, Rob Waltzer, James Salmon, and Philip Clayton. My very special thanks go to Iris Knell: amanuensis, guardian of the Robinson professoriate, and she who would never split an infinitive.

# Contents

# The Emergence of Everything

I

# The Emergence of Emergence

The writer of Ecclesiastes who proclaimed that "The thing that hath been is that which shall be; and that which is done shall be done: and there is no new thing under the sun" was taking an extremely short-term point of view. He certainly failed to reckon that the sun itself was less than 5 billion years old in a universe probably dating back sometime over 12 billion years. He did not note that the kingdom in which he lived had only been around for a few hundred years, and changes in culture and government were constantly occurring.

This book on emergence deals with ways of thinking that are new under the sun: fresh perspectives for looking at the world that are accompanying the computer revolution, a new willingness of scientists to deal with complexity, and the very construct of emergence that provides a clue as to how novelty can come to be in a very old universe. In short, we are picking arguments both with the author of Ecclesiastes and with those who think about "the end of science." Something new and exciting is taking place in analytical thought, and it promises different ways of looking at philosophy, religion, and world-view.

When I was an undergraduate, I read the philosopher José Ortega y Gasset, who explained that science is that discipline that replaces the hard questions that we are unable to answer by simpler questions to which we are competent to seek solutions. Ortega y Gasset (1883–1955) was writing about the science and mind-set of his time, when the search for simplification and mathematical certainly took precedence over approaches to complexification, thus severely limiting the domain of the sciences. The invention and elaboration of high-speed computers over the last half-century has radically changed the questions we are able to ask and has altered how we choose to pose them. Let us look at how all of this came to be.

In order to appreciate the concept of emergence and the complexity framework within which it arose, we might first take a brief tour of the history of post-Renaissance science to sense how the precomputer mindset developed and how views constantly changed. We then turn to the causes of the substantial alterations of understanding that have come with the latest findings during the past few years.

We start with what many regard as the formal beginning of modern science, the mechanics of Galileo Galilei, that focused on space and time as the appropriate variables for the study of physics. The Italian savant derived the law of falling bodies and developed the concept of inertia. These were major conceptual alterations of existing views in a time of great intellectual change. Galilei also endorsed the heliocentric theory of Copernicus. His contemporary, Johannes Kepler, was able to deduce from the observational data of Tycho Brahe that the planetary orbits were shaped like ellipses having the sun at one of the foci. Kepler also found that a line (radius vector) joining any planet to the center of the sun sweeps out equal areas in equal lengths of time. His third law of planetary motion relates the squares of the orbital periods of planets (planetary years) to the cube of the mean distance from the sun.

These observed laws of planetary motion with the sun at one focus stood as empirical generalizations until Isaac Newton formulated the laws of motion, developed differential calculus, and postulated the universal law of gravitation between any two bodies. Using these generalized laws of mechanics and gravity, it was possible from first principles to derive Kepler's laws of planetary motion. This is simple enough physics that it is usually done in contemporary undergraduate courses. I recall the thrill of deriving the laws in my own undergraduate course in physical mechanics. The cherished textbook is still on my bookshelf. These results about planetary dynamics, available since the late 1600s, were enormously powerful, because they enabled one to make predictions of planetary trajectories on the basis of mathematical law. Observations then verified the predictions. This approach established the methodological framework of physical science for the next 300 years. Note that realizing the full power of Newtonian physics required the invention of calculus by Newton and independently by Leibnitz. This mathematical advance was required in order to generate numerical solutions. The relation of mathematics to science is a matter of special interest. There are those, myself included, who believe that high-speed computation is to biology and the social sciences what calculus was to physics. Computer science is the mathematically based or formal tool that seems to map best onto the structure of the questions asked by many modern natural sciences, and moves into the domain of the social sciences.

The success of Newtonian physics had a great impact in eighteenth-century thinking in any number of fields. Alexander Pope summed up many of these ideas when he wrote:

> Nature and Nature's law lay hid in night
> God said let Newton be! and all was light.

In astronomy, mechanics, and celestial mechanics, the Newtonian approach was carried forward by the French mechanists Laplace, Lagrange, D'Alembert, and others. In optics and electricity and magnetism, the 1800s saw the work of Gauss, Faraday, Maxwell, and Hertz establish the classical branches of those parts of physics. By the end of the nineteenth century, science was firmly in place with a certain completeness in a number of areas designated as classical physics. In those domains, the mathematical postulates and boundary conditions led to firm numerical predictions that could be checked against observation. The range of soluble problems was limited only by certain severe restrictions in the mathematics.

In biology, the nineteenth century saw two grand theories, one thoroughly reductionist and one of a different character. The first was the cell theory—all living matter is made of cells, and all cells come from previously existing cells. The science of histology was developed to visualize and analyze tissues in terms of cells, and physiological chemistry began to explain cells in terms of molecules. The second theory—evolution—was enigmatic because it analyzed the appearance of all species in terms of evolution from previous taxa but had no formalism other than an unclear and somewhat tautological theory of fitness to explain which species survived. A third theory, genetics, the analysis of the hereditary transmission of traits, would have illuminated the other two, except it took 40 years to be rediscovered in the early 1900s. The original work of Gregor Mendel had never found its way into the scientific mainstream, and it took a long time before others independently discovered the same laws of inheritance.

In the late 1800s, chemistry was unified by formulating the periodic table of elements as an empirical generalization. The picture was confused by debates about whether atoms were real or simply explanatory devices. This intense battle has, I believe, disappeared—for all theories consist of explanatory devices to predict phenomena, and "real" atoms are equivalent to the unknowable *"ding an sich"* (thing in itself) discussed by Immanuel Kant that underlies the phenomena. It is a symbolic argument between the positivists and the realists. The question doesn't have to be answered in order to proceed, but the argument persists.

Many of these issues regarding atomism came together in the life and suicide of the great Austrian physicist Ludwig Boltzmann in 1906. His biographer E. Broda wrote:

> A factor contributing to his death may have been his feeling that the atomic theory, for which he had fought throughout his life, was being pushed into the background. This opinion was expressed, for example, by his Leipzig student, Georg Jaffe. The influential Alois Höfler, a personal friend but philosophical opponent, wrote after Boltzmann's death in 1906: "The enemies of traditional atomism who were led by Ernst Mach liked to call him [Boltzmann] the 'last pillar' of that bold mental structure. Some even ascribed his symptoms of melancholia, which went back for years, to the fact that he saw the tottering of that structure and could not prevent it with all his mathematical skill.
>
> . . . It was tragic that opposition to the atomic theory contributed to Boltzmann's depressions, for it was precisely at the time of his death that the atomic theory achieved its greatest victories. Des Coudres wrote: "Here Boltzmann deceived himself to his own detriment. . . . And also the banner under which our young experimenters make their surprising discoveries—be it the ultramicroscope, the Doppler effect in canal rays, or the wonders of the radioactive substances—is the banner of atomism; it is the banner of Ludwig Boltzmann." By 1906 atomism had already weathered the period of lowest repute, thanks in large measure to the new experimental results.

These new experimental results were gathered together in 1913 in an extraordinary work, *Les Atomes,* by Jean Perrin. In perhaps the greatest triumph of connecting different approaches in classical science, Perrin focused on determining Avogadro's number, the presumably universal number of molecules in a gram molecular weight of a substance. Perrin reviewed 16 very diverse methods of determining the number, many of which he carried out experimentally in his own laboratory (See Table 1, below).

The methods chosen by Perrin are a mirror of the physics and chemistry

TABLE I: VALUES OF AVOGADRO'S NUMBER

| Phenomena Observed | $N/10^{22}$ (Avogadro's number) |
|---|---|
| Viscosity of gases (kinetic theory) | 62 |
| Vertical distribution in dilute emulsions | 68 |
| Vertical distribution in concentrated emulsions | 60 |
| Brownian displacement | 64 |
| Brownian movement: Rotations | 65 |
| Diffusion | 69 |
| Density fluctuation in concentrated emulsions | 60 |
| Critical opalescence | 65 |
| Blueness of the sky | 65 |
| Diffusion of light in argon | 69 |
| Black body spectrum | 61 |
| Charge as microscopic particles | 61 |
| Radioactivity: Projected charges | 62 |
| Helium produced | 66 |
| Radium lost | 64 |
| Energy radiated | 60 |

of his time. The viscosity of gases can be calculated from the kinetic theory of gases and depends on the number of molecules per unit volume. Since this quantity is the number of moles per unit volume times Avogadro's number, the experimental value of viscosity yields the desired quantity.

The distribution of emulsions in a gravitational field is calculated by statistical mechanics and depends on the potential energy of the particles mgh (mass times gravitational acceleration times height) divided by the kinetic energy kT (Boltzmann's constant times the absolute temperature). Since Boltzmann's constant is the gas constant available from Boyle's law divided by Avogadro's number, its value determines the desired quantity.

The next three methods depend on measuring Brownian motion, the random migration of microscopic particles in a gas or liquid. (Robert Brown first observed this for pollen grains in water.) In 1905 Einstein developed a theory to explain this phenomenon that was based on molecular kinetic theory of liquids. By observing the trajectory of Brownian particles, Perrin was able to calculate the Boltzmann constant and hence Avogadro's number.

The next four methods are based on light scattering that is due to local fluctation of the number of molecules per unit volume. This leads to local fluctuation in the index of refraction and light scattering. (Among other things, this is responsible for the blue of the sky.) The fluctuation depends on the number of molecules per unit volume in the gas that is the number of moles per unit volume times Avogadro's number.

For a novel determination, Perrin went to Planck's famous 1901 formula for the spectral distribution of black-body radiation. It can be fit by two constants, **h,** the Planck constant, and **k,** Boltzmann's constant. The latter is, as noted, the gas constant divided by Avogadro's number, leading to an independent determination.

The next method is based on electrochemistry, where the charge per gram molecular weight of univalent ions has been determined and called the Faraday. With the first determination of the unit electron charge, it was clear that the Faraday divided by the charge on the electron was Avogadro's number.

For the last four values, Perrin turned to the newly discovered phenomenon of radioactivity from which he found four methods of determining the universal constant of Avogadro. One of these illustrates the phenomenon. In $\alpha$ particle decay, an ionized helium nucleus is ejected. The number of decays can be counted with a scintillation counter, the helium can be collected as helium gas, and the amount determined volumetrically. Thus,

$$\text{\# decays/Avogadro's number} = \text{moles of helium.}$$

By counting decays and determining moles of helium, Avogadro's number can be directly determined experimentally.

Perrin concluded:

> Our wonder is aroused at the very remarkable agreement found between values derived from the consideration of such widely different phenomena. Seeing that not only is the same magnitude obtained by each method when the conditions under which it is applied are varied as much as possible, but that the numbers thus established also agree among themselves, without discrepancy, for all the methods employed, the real existence of the molecule is given a probability bordering on certainty.
>
> The atomic theory has triumphed. Its opponents, which until recently were numerous, have been convinced and have abandoned one after the other the skeptical position that was for a long time legitimate and no doubt useful. Equilibrium between the instincts towards caution and towards boldness is necessary to the slow progress of human science; the conflict between them will henceforth be waged in other realms of thought.

Note that in all 16 cases, theory permitted one to carry experiments that gave rise to the numbers. The numerical agreement was the verification that validated the theories. This is a key feature to acceptance of a theory in classical physics.

The atomic theory is central to physics, chemistry, and biology. Just at the time Perrin was doing the experiments leading to *Les Atomes,* Einstein and Planck were doing the work that gave us relativity and quantum mechanics and a whole new view of the physical world. Bohr was simultaneously formulating the theory of the energy levels of atoms. Before going on to this new world, let's review where science stood at the beginning of the twentieth century.

Mechanics, electricity and magnetism, optics, hydrodynamics, thermodynamics, kinetic theory, and celestial mechanics were solidly established as the firm foundation of physics. Biology was beginning the great age of genetics, and physiology was searching for its chemical roots. Organic chemistry was finding explanation in the tetrahedral geometry of carbon bonds, and organic synthesis was being extended to a wide variety of new products. Discoveries of the structure of sugars, amino acids, and nitrogen heterocycles were providing a firm basis for biochemistry. A general but not universal agreement was beginning to arise that biology could be reduced to chemistry, which could be reduced to physics.

The often unstated philosophy of science was based in its various forms on starting with observation, developing theoretical explanations of the observations, and using these to predict other observations. The success or failure of the predictions provided the epistemological roots of any science. The paradigm example of this kind of science was the study of the solar system, where future trajectories of planets could be predicted with great accuracy. The social and cognitive disciplines were viewed in a totally different domain than the physical and chemical sciences. Biology stood between them, looking in one direction toward chemistry and in the other toward ethology and anthropology.

There were some attempts to bridge the gaps. Economists in the late 1800s had discovered thermodynamics and were attempting to use the mathematics of that science as a framework to develop theory; however, the approach lacked the predictive power of physics.

The general approach to the philosophy of science followed through for the twentieth century. Two books outline the general approach: *The Logic of Scientific Discovery* (Karl Popper, 1934) and *The Nature of Physical Reality* (Henry Margenau, 1950). Popper provided a prescriptive approach for the logical requirements for a subject to be an empirical science. Margenau provided a descriptive approach of the intuitive metaphysical assumptions that physicists make in formulating and accepting a physical theory.

Both approaches start with the observable world and move to the formulation of a theory, usually mathematical, to explain relations among observables. The theory then makes predictions about other observables and is tested by comparing predictions with observations. The theory stands or falls on the agreement or disagreement, usually numerical, between prediction and observation. It is difficult to put the theory of evolution in this context, so that biological science did not fit this epistemic scheme nearly so well as physics, a failing constantly stressed by the ideological enemies of "evolution." When we discuss the emergent nature of evolution, it will be clearer why biological science did not fit the simplified scheme.

Both Popper and Margenau dealt with the subject of the epistemology of science: "How do we know?" This kind of inquiry had been established by Immanuel Kant in his critiques. It has not been a popular subject in science, and less so in religion where knowledge by faith is the ultimate test. I consider epistemology crucial to our understanding.

In science we start with the immediately given, the sense data that are of course the contents of minds. From these sense data that are shapes, colors, sounds, feels, and meter readings, we develop theoretical constructs such as solid objects, atoms, electrons, and probability waves. The constructs, as Kant points out, are not the incompletely knowable "thing in itself," but deal with the contents of our minds. Science starts with the mind, both the perceiver of sensations and the postulator of constructs. Science also assumes a community of minds who can agree on the sense data and the verifiability of consequences of the constructs. Regardless of one's philosophical position, science begins with the mind and is a public activity. Constructs have a hierarchy from quarks to atoms to molecules to organisms. The contemporary position of most neurobiologists is to try to go up the hierarchy from atoms to minds to understand the emergence of mind in terms of the underlying members of the hierarchy.

This of course presents an epistemic circle. One starts with mind as the primitive and goes around the circle of constructs in an effort to explain mind. I have no trouble with this circularity, but it comes as a surprise to many scientists. It is an epistemology that somehow accords with the emergence view of the evolving universe, or at least our part of it.

In terms of this view, one can understand materialists or naïve realists as individuals who believe that the constructs of particles are more real than the minds that constructed them. Idealists in the philosophical sense are individuals who believe that the minds are more real than the hierarchy of things that constitute the world out there, the things in themselves. I find both views much less enlightening than accepting the circle as the ontological sequel of this type of epistemology. It recognizes the existence of the world out there without requiring people, but also recognizes that the kind of knowledge we have of that world is not independent of us, and we will never have God's knowledge of the thing in itself.

In my series of hierarchical emergences, I operate without commitment to an ontology, which may be unknown, but I do adopt the epistemology that has made physics work. However, in understanding the new views of emergence, we will find this epistemology will require some developments that have not yet been discussed.

A sharp distinction is often drawn between the immediately given sensory inputs and the rational constructs. These distinctions are quite fuzzy, and the mind operates with both, often without a sharp distinction so that observations already have a theoretical component, and constructs are often not far from the immediately given. This need not cause philosophical problems; the world is what the world is. The clear distinction between mind and nature simply does not exist.

Two developments in physics at the turn of the century were harbingers of ideas whose full philosophical significance would not be generally appreciated for almost 100 years. The central concepts of emergence trace back to the statistical mechanics of Ludwig Boltzmann, James Clark Maxwell, and Josiah Willard Gibbs. The main idea of deterministic chaos was formalized in the work of Henri Poincaré on the stability of the solar system.

The founders of the statistical mechanics assumed the atomic molecular view of matters and further posited that the atoms and molecules obeyed the laws of mechanics. They were then interested in showing how the macroscopic laws of thermodynamics and kinetic theory could be obtained from the mechanics of the reductionist agents, the atoms and molecules. By dealing with ensembles of particles or ensembles of states and showing that the macroscopic observables were averages over microscopic states, they were able to deal with variables like pressure and temperature as emergent properties. Thus while Perrin and others were pursuing the development of the reductionist view of atoms and molecules as the operative agents, the statistical mechanicians were showing that the microscopic particle view led to the macroscopic laws of thermodynamics in terms of emergent properties. This is a model that we should keep in mind in going back and forth between reductionism and emergence in the study of hierarchical levels.

Thus, while statistical mechanics has some features similar to modern emergence theory, in one very important way it is totally different. In the Gibbsean approach, one assumes that the time average of a behavior of a simple system is equal to the average of a whole ensemble of possible entities chosen to represent the system of interest; thus, the pruning rules force the behavior to converge about the mean, rather than the divergence that sometimes occurs in the nonequilibrium systems we study in contemporary examples of emergence. The solution to the seeming paradox is that the classical case deals with the unique state of equilibrium, which is a global extremum and sits at the bottom of an energy well. Complex systems are generally far from equilibrium and are represented mathematically by rugged landscapes in a phase space. There is a radical difference between equilibrium and nonequilibrium systems. The latter cannot be treated by global extrema, a mistake often made by those who haven't focused on how different the two cases are and assume they can derive biological behavior as extrema.

Henri Poincaré was a French mathematician in the tradition, going back to Isaac Newton, of the mathematical study of the workings of the solar system, the orbits of planets, and more detailed considerations. When we celebrated the triumph of the law of Newtonian mechanics and gravity predicting Kepler's laws of planetary motion, we ignored a problem in the approach that Poincaré subsequently considered.

The Keplerian laws and the Newtonian explanation came from dealing with only two bodies, the Earth and the sun. When later theoreticians tried to include the moon and the other planets in the calculation, they discovered a severe problem. For systems of three or more bodies, exact analytical solutions to problems in mechanics were not possible. The difficulty was deep within the mathematics employed. Following Newton, generations of mathematicians tried unsuccessfully to solve the three-body problem analytically, and they all failed.

A parallel difficulty was seen in the study of the stability of the solar system. Were the orbits of the planets fixed for all time, or would they change in some unknown way? In the late 1800s Poincaré undertook the problem and discovered certain uncertainties in celestial dynamics that we would now designate as deterministic chaos. It was not possible to predict the orbits for all time. One hundred years later, Poincaré's finding became central to chaos and complexity theory.

The physics of the nineteenth century viewed the scientist as an observer removed from the operation of the system under study. This changed in three ways in the first half of the twentieth century. First, the special theory of relativity referred all measurement to the frame of reference of the inertial system of the *observer,* thus more closely relating the scientist to the system of study. Second, one view of quantum mechanics reduced a probability distribution function to an event when an observation was made by a classical observer. This made the scientist as observer a necessary part of the system under study. Lastly, information theory identified entropy with a measure of the observer's ignorance of which microstate a given system was in when the macroscopic state was known. The probabilistic nature of quantum mechanics fuzzed out the firm nature of physical reality that characterized classical physics. All of this was nevertheless consistent with the epistemic loop from observation to theory to observation that characterized most of reductionist science, but established a special role for the observer.

Biology, which began the twentieth century as an observational science to classify organisms and place them on an evolutionary tree, became over the next century the most reductionist, atomistic, and structural discipline of all the sciences. Molecular biologists reduced all process to the operation of known chemical structures. Molecular biology, symbolized by the double helical structure of DNA, achieved enormous success, the ultimate in what one could achieve with this approach to science. Only when one got to neural or cognitive science was it necessary to return to the problem of the observer in biology.

An example of what one can and cannot do in the context of reductionist molecular biology is helpful. If we have a purified protein, we can cause it to form into crystals, and by X-ray diffraction we can determine the precise three-dimensional structure, atom by atom. Now suppose we have the amino acid sequence of a protein derived from knowing the DNA sequence of the gene that codes for it. We wish to calculate structure from sequence. Assume we have all the interaction energies as a function of distance between various amino acids, and we wish to calculate the configuration of minimum energy. There are so many possible configurations that a computer the size and age of the universe cannot enumerate all the possibilities. Such calculations we designate as transcomputable.

We need ways of doing or short-circuiting such a calculation by selecting or pruning or radically eliminating most of the states. The emergent solution gives some idea of the route to the folded state. Selection algorithms are required to reduce the dimensionality of the problem to something that can be comprehended.

The attempt to calculate the answer over the entire domain of states is the conventional Popperian (*The Methodology of Karl Popper*) approach to protein folding. Introducing selection algorithms to look for plausible solutions is an epistemological approach that is quite different and much more difficult to evaluate by falsification, and it is too easy to be impressed by metaphoric verification. Selection is, however, an approach to using science for a great variety of problems not previously amenable to study. Where it works, novelty may appear and new ways of looking at the world may emerge. We will return to this approach after examining a number of emergences that have made the world what it is.

At the end of the nineteenth century, a number of branches of the science of biology were maturing into the discipline of neurobiology. The work of Ramon y Cajal and Camillo Golgi established the histological foundations of the nervous system as an enormous collection of neurons, cells connected into an exquisite network of vast complexity. The human brain was found to be composed of the order of magnitude of ten trillion nerve cells. Though sensation and cognition are all related to brain activities, there is a gap between the physiological activities and the simplest act of thought. Though we understand the histology of the nervous system and the physiology of action potentials and synaptic activity, we are far from dealing with the nature of consciousness and other integrated properties of the nervous system.

At mid-twentieth century, when computers began to deal with information on a grand scale, new questions were raised. John von Neumann, one of the founders of computational science, wrote an extraordinary book in 1957, in which he raised the questions associated with the relation of *The Computer and the Brain.* Branches of computer science such as neural nets have been part of attempts to model algorithmically the activities of collections of neurons.

All of the developments in neurobiology are part of an ongoing effort to understand how thought emerges from the activities of organisms. I think it clear that we are only at the very beginning of this most important quest.

Throughout the twentieth century there was an increasing awareness that biology dealt not only with matter and energy, but also with information. When information became formalized in the work of Claude Shannon, biologists immediately responded, and in 1953 Henry Quastler edited *Essays on the Use of Information Theory in Biology.* Genetics and evolution have adapted the language of information theory, which also finds expression in linguistics. Biology at the molecular or global ecological level is information-dense, and this has provided a background of a branch of mathematical biology. The biological emergences that will be discussed have a component of information emergences.

The main feature that has characterized theoretical science of the last third of the twentieth century is the development and use of the high-speed digital computer as the major tool. An entire series of problems of physics and chemistry that were conceptually understood had been set aside as intractable by the available analytical techniques of mathematics. Using computational techniques for some such problems had been practiced for a long time without mechanical aids, going back to famous mathematicians such as Carl Friedrich Gauss. Computers increased the speed of analysis and the complexity of problems that could be so treated by many, many orders of magnitude.

As a result of the great success of computation, increasingly more complex problems in physics and chemistry were undertaken, and scientists were led to computer modeling in branches of biology and social sciences such as economics. The new areas and approaches were brought together under the heading of the sciences of complexity, and common features among the disciplines were sought.

In a number of problems being modeled, beginning with Edward Lorenz's work in meteorology, the computer trajectories were exquisitely dependent on the starting conditions or boundary conditions. This resulted in deterministic chaos where the result could not be calculated because the boundary conditions could not be known with sufficient precision. The computer, limited by the number of significant figures it could carry, could not calculate with sufficient exactness to get an answer, or even an approximate answer. The entire field of deterministic chaos developed.

In another class of problems, the interaction rules are known, but the complexity is so great (or number of possible states is so great) that the problem surpasses the capacity of any known computer or even any conceivable computer. Rather than give up on such problems, scientists have looked for ways of pruning the space of possible solutions or sets of allowable solutions. This may lead to surprises in the system trajectories, giving rise to novel behaviors. These are the emergent properties of the system, properties of the whole. They are novelties that follow from the system rules but cannot be predicted from properties of the components that make up the system. The individuals that make up the whole are designated agents. For example, interaction rules of individual insects (the agents) may give rise to the configuration and behavior of swarms (the agents at the next hierarchical level).

The reductionist approach leads us continually to seek solutions at lower and lower hierarchical levels. To move conceptually in the other direction, we must apply pruning algorithms and seek for emergent properties or entities that become the agents for advancing another hierarchical level.

Emergence is both a property of computer models and of the systems being modeled. And so nature yields at every level novel structures and behaviors selected from the huge domain of the possible by pruning, which extracts the actual from the possible. The pruning rules are the least understood aspect of this approach to emergence, and understanding them will be a major feature of the science of the future. Clearly, new epistemological approaches will be required. A kind of framing-up of science to enter novel domains has been accomplished.

Emergence is then the opposite of reduction. The latter tries to move from the whole to the parts. It has been enormously successful. The former tries to generate the properties of the whole from an understanding of the parts. Both approaches can be mutually self-consistent.

## Chapter 1—Readings

Broda, Engelbert, 1983, *Ludwig Boltzmann,* Ox Bow Press.

Gell-Mann, Murray, 1994, *The Quark and the Jaguar,* W. A. Freeman and Company.

Holland, John H., 1995, *Hidden Order: How Adaptation Builds Complexity,* Addison Wesley.

Kauffman, Stuart, 1995, *At Home in the Universe,* Oxford University Press.

Lindsay, Robert, 1933, *Physical Mechanics,* D. Van Nostrand Co., Inc.

Margenau, Henry, 1950, *The Nature of Physical Reality,* McGraw Hill. (1977 Reprint, Ox Bow Press)

Perrin, Jean, 1913, *Les Atomes,* Librairie Félix, Alcan (English translation: *Atoms,* 1990, Ox Bow Press).

Popper, Karl, 1959, *The Logic of Scientific Discovery,* Basic Books.

Quastler, Henry, 1953, *Essays on the Use of Information Theory in Biology,* University of Illinois Press.

Von Neumann, John, 1954, *The Computer and the Brain,* Yale University Press.

# 2

# Ideas of Emergence

In the beginning the universe was unimaginably hot and inconceivably dense. It was chaotic and void of matter as we know it. Within this exploding mystery of mysteries lay the roots of all subsequent existence, the next 12 billion or so years from the original creative event for our universe to the much more mundane but scientifically still mysterious act of my taking pen to paper to comment on exploring the unfolding of the universe in time and the meaning of that process to me and my fellow inhabitants of our Earth. Both the workings of the cosmos and the activities of the human mind call out for our understanding. Contemporary scholars in many disciplines are impelled to answer the call. Emergence is a new and promising tool in that understanding.

The probing of the universe that we are here undertaking exemplifies a kind of scientific passion to understand the big picture that goes back, at least, to the Roman scholar Lucretius and his poem, *"De Rerum Natura"* ("On the Nature of Things"). More than 1,500 years following Lucretius, the rise of early modern science from Galilei and Newton (1600–1800) through Maxwell and Darwin (1800–1900) and a host of others has provided a great wealth of knowledge, with constant addition of new material for attempting this kind of universal in-depth examination. In the 1940s, Pierre Teilhard de Chardin, a Jesuit paleontologist who is my role model in this kind of speculative scholarship, attempted, in *The Phenomenon of Man,* to examine all of cosmic history in evolutionary perspective from the origin of the universe to the origin of the human spirit and beyond. He struggled with all his being to reconcile his science and his religion. The twentieth century has seen this kind of search for God within the laboratory and in the field as a recurring and expanding theme. My biological colleagues have had little sympathy with Teilhard, but I ask them for a rereading and sympathy for his attempt to seek for something deeper in the evolutionary unfolding of our universe.

At the end of the second millennium of the Common era, which has concluded with the most dynamic and creative century in the entire history of science, we now see the world through the fresh perspective and understanding of the computer revolution and the study of complex systems. I entered college in the 1940s, and I have seen the vast changes unfold before my eyes. I'm not sure that all contemporary scientists and policy-makers realize just how profound the changes in perspective have been during the past century. I believe that this conceptual revolution is comparable in importance to the discovery of language or the discovery of mathematics. In the last few years, this new mode of thinking has begun to develop an exciting explanatory concept designated *emergence,* which develops previously unrealized ways of deepening our understanding of the past eons and illuminates how the universe, after a long and complex 12-billion-year trajectory from the Big Bang, has given rise to the human mind and modern man. The task of the following pages is to look at a number of specific emergences and to demonstrate how the concept of emergence itself, with all its uncertainties, helps inform our comprehension of the unfolding of cosmic history. When we have examined more fully the evolution of the universe in the modern context, the results of the probing will, I believe, like the "flower in the crannied wall," have much to tell us about "what God and man are."

In spite of the recent revolution, the philosophical, theological, and scientific thoughts that we discuss in these pages have not appeared suddenly at a given instant in history. They have been formulated and refined by generations of thinkers throughout the ages. Some of the main ideas go back at least to the Golden Age of Greek philosophical thought, and others originated in the search for purpose in history that was developed and formalized by a group of wandering Semites in the deserts of the Mideast. In truth, it is reasonable to assume that some ideas we build on must go back even further to the time when our hunter-gatherer ancestors first developed agriculture and settled down for a long enough period of stability to think about the big questions. Thus, in the cultural milieu of Western civilization, one major stream of our ideas arose in the age from Thales to Theophrastus when the foundations of Western philosophy were laid down within the academies of Hellenic civilization. The second stream of thought from a nearby part of the Mediterranean world originated in the epic from Moses to Ezra and Nehemiah and is embodied in the volumes of the Old Testament and commented on in Talmudic literature. A third view of God arose among certain followers of Jesus of Nazareth. This God of Faith, perhaps best exemplified in the writings of Paul of Tarsus, is known by epistemological processes of a totally different type, related to but not derivable from the previous ways of knowing. These historical strands have generated a vast 2,000-year-old effort to understand three views of the deity: the God of Faith, the God of Reason, and the God of History. Modern efforts are part of an ongoing attempt at that understanding. Scholars of the Middle Ages struggled passionately with these questions. Fortunately for us, many of the major works in this quest have been preserved and are available for study.

In rereading the preceding paragraph, I am struck by the extent to which I have focused on Western science and the Abrahamic religions and have ignored the thought of the Orient, the Indian subcontinent, Africa, and the Americas. In part, this is because science, as we know it, is largely a product of the Western world and has been in constant contact with the philosophy and religion of the residents of that part of the planet. It also stems from my ignorance of the vast amount of thought of the civilizations outside of the one in which my education occurred. *Mea culpa;* we all have our limitations. Emergence also is most easily seen within that world of Western science. I am, nevertheless, sure that I am missing things to be found in the ignored traditions.

The two-millennia-old trichotomy of views about God's nature now can be reexamined within the perspective of the new approaches of complexity theory and the constructs of emergence that themselves challenge the fundamental epistemological approaches to sciences that had come to maturity in the 1930s and 1940s and are exemplified in the works of philosophers of science Karl Popper and Henry Margenau, which we discussed in the last chapter. Following these philosophical works, and with a powerful assist from high-speed computers, scientists have tried to extend the conceptual approaches of physics and chemistry into the realms of biology, economics, the psychological and social sciences, and all manner of applications. This has led to new ways of looking at the world and even to new views about the very task and the possibilities of science. In the end, these novel approaches are serving to energize and enhance the conversation between science and the humanities and between science and religion.

Three vastly different time scales enter into our considerations of the science and philosophy of emergence. First are the past 30 or 40 years, when researchers have moved from a solidly reductionist approach to science to a much looser, complexity-oriented view of the task of theory. The second time domain is the more than 10,000 years of human civilization that have produced our ideas of God and man. We must constantly remind ourselves what a short period of time this is and of the novelty of human civilization. The third and much vaster domain in time is the 12-billion-year unfolding of the history of the universe from the unfathomable beginnings to the mind of modern humans who attempt to turn back on the history of this universe and ask, "What does it all mean?" We will deal with all three of these chronometric domains.

Let us then set the stage for expounding the nature of emergence with a brief view of the philosophy of science before the computer revolution. All science within the classical epistemological framework, which goes back at least to Descartes, starts with the observed world of the senses: the sights, sounds, smells, tastes, and touches that make up our sensory perceptual experiences. Science as we now practice it begins with this phenomenological domain. Immanuel Kant (1724–1804), the dominant classical philosopher of science whose views are still central and worth considering, has reminded us that perception is more than just pure sensory experiences, and we must always keep in mind the mentally imposed features of our world-view that we bring to our observations (the *a priori*). Beyond this mostly sensory world, we move, as noted in the last chapter, into theory or construct domain in which the colors, shapes, and tactile sensations are interpreted as objects. The green, hard, cylindrical shape achieves more permanence as the pen with which I write these lines. Science tends to deal mainly with shared public perceptions that we can agree on universally and at first avoids dreams, visions, and other experiences available only to lone individuals. Individual religious experiences are also excluded because of their private nature.

Classical science then moves by consensus through a series of abstractions such as material objects→ molecules→ atoms→ electrons→ probability distribution functions, and even more abstract notions. This pathway is known as reductionism and theory formation. From the theoretical constructs postulated at each level, we can make a series of predictions or rules that work their way, often through calculations, back to the world of observation. The results of theory prediction and observation can then be compared. If they agree, particularly numerically, the theories are tentatively verified, and if they disagree the theories are falsified and must be discarded or altered. Reductionism in physics is the explanation of what goes on at one level in terms of theoretical constructs at another level one or more stages further removed from observation. In biology the hierarchy from the construct of organisms to organs to tissues to cells to organelles to macromolecules has a somewhat more empirical character than theory formation in physics, but the use of reductionism as explanation in terms of more abstract levels and smaller entities still obtains. In the most mature sciences we seek for exact numerical agreement between measurement and theory. Science in this view consists of those theories that have not been falsified. It is constantly subject to change.

In chemistry, in classical and quantum physics, and in molecular biology, reductionism has been extraordinarily successful. Using the epistemology based on falsification and verification as elaborated by Karl Popper, it has been possible to unify a vast array of experimental findings. The spectacular success of reductionism emboldens us to ask questions that lie, because of their scope, outside the usual paradigm. Examples are: How did life originate? How does evolution produce new species? How do the actions of individuals produce societies, economies, and markets? We also inquire about the origin of novelties such as human minds and corporations. In general, these kinds of questions force us to go in a direction opposite to that of classical reductionism. We take the knowledge and rules at one hierarchical level and try to apply them to derive the rules and structures of the higher hierarchical levels.

The reductionist description will generally postulate entities called agents, often a large number of such agents, usually interacting by nonlinear rules. The reactions cannot be described by simple relations such as A is directly proportional to B. In the precomputer age, the approach to prediction was to formulate complex problems in terms of equations, usually a number of second-order nonlinear differential equations, and then stare at the insoluble series of relations, searching for linear approximations. The equations represent the rules for the interaction of agents. Using computers, one is able to move directly to numerical solutions. But a new general feature is that the solution space becomes huge and highly multidimensional, and the problems become combinatorically explosive. The next and critical step in the computational approach to such problems has been to select solutions or families of solutions by fitness rules or other selection criteria often defined by introducing pruning algorithms. Theories of this kind are successful if the selected set of solutions or the solutions generated under the constraint of rules and pruning lead to behaviors with some kind of agreement or resonance with the world of observation. Such outputs are called emergent properties of the system. The agreement between computation and observation by these approaches is not necessarily numerical and exact, as it is in classical science, but it does lend insight into the problems being studied. Indeed, the notion of metaphor sometimes replaces agreement, an idea that worries some scientists, but may allow us to move ahead.

In the domain of emergence, the assumption is made that both actual systems as well as models operate by selection from the immense space and variability of the world of the possible, and in carrying out this selection, new and unanticipated properties emerge. This type of outcome is similar in some ways to the biologists' view of evolution, in which novelty occurs by mutation, translocation, selection, and differential survival. New structures, new species, and new ecosystems thus emerge. The evolving taxa and systems are not predictable in any exact sense. Thus, emergence has a certain familiar feel to biologists.

Emergences thus occur both in model systems and in real world situations. If the models are well chosen, the two kinds of emergences map onto each other. They resonate with each other. In both cases, emergence leads to novelties: the whole is somehow different from the sum of the parts. The outcome cannot be known without running the computer program. With respect to philosophy of knowledge, this kind of selection or pruning is a new construct, and we don't yet have a firm sense of validating or rejecting theories, as was the case in more traditional philosophy of science. We lack well-defined epistemological criteria for this new type of approach. In spite of these problems that result from our lack of epistemic understanding, the real gain is that both reduction and novelty can exist together in the same framework, and we can follow the evolutionary unfolding of the world in terms of significant emergences, even though we lack exact predictions and precise understandings of what is taking place. A different way of doing science appears before us, and we are going to have to develop a philosophical framework for this new sense of reality. This is not to be construed as a barrier to moving ahead; in classical physics we used what was essentially the Popperian approach to scientific epistemology long before it had been made explicit. Emergence does not mean randomness; it is an orderly unfolding of the world, but an unfolding rich in novelty. We know the challenge, if not the solution.

Emergence as a concept in evolutionary biology became a major view with C. Lloyd Morgan's 1923 book *Emergent Evolution*. Variety and novelty were introduced as irreducible features of the temporal unfolding of evolution. Increasing complexity is a consequence of these properties and leads to hierarchical levels that Morgan notes as physics, life, mind, and spirit. The discussion of the ideas of the early emergent evolutionists is quaint and at the same time resonant with ideas that appear in current works on the subject.

As early as the 1960s, a new wave of thoughts about emergence occurred. Henry Quastler's 1964 book *The Emergence of Biological Organization* dealt with the problem of how biological information arose in a chemical system. Walter Elsasser, in his 1969 book *The Atom and the Organism,* dealt with the dilemma that biology cannot be predicted from quantum mechanics because of the enormous complexity in even the simplest cell. He hinted that biological laws come about from a selection or pruning, but being far ahead of his time, he lacked the contemporary vocabulary and computer analogues for his ideas.

John Holland has described the contemporary view of emergence coming from computational science. He gives as an example a checkers-playing program that has strategy rules and is capable of learning—that is, of altering the rules and parameters in response to playing against human players or other programs. After a lot of game playing, the program can defeat its designer at checkers, but its emergent strategy is opaque to the designer. A novelty emerges, unpredictable because of its complexity. All possible pathways have been pruned to a subset by experience. The pruning generates a fitness for winning.

The checkers example can be generalized to cases in which there are agents (checkers) and interaction rules (the rules of the game plus the strategy rules). Pruning algorithms generate the strategy changes that are calculated from the results of playing a large number of games. A new strategy that may represent genuine novelty emerges.

Computational modeling is new because many agents and nonlinear interactions lead to a combinatoric explosion of possibilities, and pruning (or selection) is necessary to get any understandable solution. We do not have a theory of pruning or an epistemology that is appropriate to this kind of modeling. Nevertheless, it is the kind of theory necessary to the hierarchical ladder of complexity from the most basic constructs to more complex arrays.

To begin reviewing emergences in more social and religious contexts as we shall wish to do, a brief discussion of some related philosophical ideas is in order. First, there are a series of scientific-theological approaches from Giordano Bruno and Benedict Spinoza to Albert Einstein that identify God with the laws of nature. These ideas (to be elaborated later) can probably be traced back to Philo of Alexandria and the Athenian School of Plato and Aristotle. This God of natural law or the God of Reason is an immanent God, mysterious and probably impenetrable or unknowable. This pantheistic view provides a context within which the study of the laws of nature rises to a religious undertaking, an attempt to know the mind of God. In such a context, science is a vocation. However, this God of the laboratory and observatory is cold and distant and listens to no one's prayers. But according to Spinoza, he may and should be adored. To some this seems a contradiction. Lacking the emotional impact of traditional Western religion, it seems quite unsatisfactory for the religion of the people.

In the mid-twentieth century, Pierre Teilhard de Chardin built a world-view based on the newly discovered evolutionary ideas of Charles Darwin and the philosophical ideas of evolution as elaborated by Henri Bergson and C. Lloyd Morgan, in which the unfolding of the laws of nature has a direction in time. Though the laws are timeless, if there is a beginning, then the state the system evolves to is subject to its laws. *The Origin of Species, Creative Evolution, Emergent Evolution,* and *The Phenomenon of Man* are a collection of evolutionary works that, while philosophically disparate, share a directionality in viewing the history of the cosmos. There is thus a teleological-like feature of trying to understand God's mind that scientists have perhaps been too hasty to reject. After all, we start with observations, and if the evolving cosmos has an observed direction, rejecting that view is clearly nonempirical. There need not necessarily be a knowable end point, but there may be an arrow.

That there was a beginning has been elegantly spelled out within the constructs of contemporary science in Steven Weinberg's concise book *The First Three Minutes,* and further developed in the Grand Unified Theory. These works take us back to the time beyond which we cannot penetrate with current normative constructs of theoretical physics. In *The Life of the Cosmos,* astrophysicist Lee Smolin tries to go back at least one step further, to the birth of universes within black holes. We will have to wait to see where this kind of thinking will go. As a matter of taste, I don't like to focus on the limits of thought, as do some of my colleagues. We are only at the beginning of understanding and should be wary of narrowing our possibilities prematurely. On the other hand, I am content to begin with the Big Bang. It satisfies my curiosity, since I am trying to puzzle out the origin of life that is clearly a post–Big Bang phenomenon.

How do changes within a system occur, given the framework of unchanging laws? At this level, our newfound complexity theory introduces something radically different into the discussion of change. Because of complexity, reductionist rules operating at one level can produce emergent unpredicted consequences or rules at a higher level. These consequences in no way violate the underlying laws, but follow from these principles, subject to the selection or pruning rules. The selection criteria may be deterministic or may be fitness rules or may have other roots such as frozen accidents, but they represent possible devices for influencing the unfolding of the reductionist systems in time. Frozen accidents are random events that so alter the system that their effects persist. An example would be a genetic mutation triggered by a cosmic ray passing through a gene. The higher level in the hierarchy may show novelty. There are many kinds of selection rules, but they often may lead to a whole that is different from the sum of the parts, i.e., the behavior of the agents leads to system properties not knowable without running the program. This is a recurring theme.

This idea of parts and wholes is seen in statements from radically different perspectives by Pope John Paul II and scientist John H. Holland. In 1992 the Pope, in giving the allocution to the plenary session of the Pontifical Academy of Science on *The Emergence of Complexity in Mathematics, Physics, Chemistry and Biology,* asked, "How are we to reconcile the explanation of the world . . . with the recognition that the whole is more than the sum of its parts?" John H. Holland, writing in his book *Emergence: From Chaos to Order,* notes: "Emergence, in the sense used here, occurs only when the activities of the parts do *not* simply seem to give the activity of the whole. For emergence, the whole is indeed more than the sum of the parts." Curiously, the identical words "the whole is more than the sum of the parts" occur in the scientific treatise and papal discourse. This seems to speak volumes to the range of the ideas of emergence and its possible role in the science/religion dialog.

The feature of novelty is an aspect of emergence we wish to build into our thinking. It is a new concept; one not completely understood. Indeed, in some aspects it is only dimly comprehended. An insightful discussion of emergence in the book by John Holland we have just cited stresses the incompleteness of our understanding. In the next chapter, we will proceed by presenting in outline some 28 examples of emergence, which, while different in character, form a sequence in time from the earliest beginnings of the universe to the future of mankind. Subsequently, we discuss the 28 examples in more detail. These case studies add depth to the ideas of emergence sketched out in Chapters 1, 2, and 3 and elaborated in the following chapters. One such view discussed below can make contact with the religious philosophies, as suggested above. If we identify the immanent God, the mysterious laws of nature, with God the father, then emergence will be the efficient operation of that God, which Christianity views as the Holy Spirit. We will come back to these ideas in more detail later. Twelve billion years downstream when *Homo sapiens* emerges as the result of the operation of the natural laws, we may think about God in human affairs, which resonates with the Trinitarian ideas of the Son or man in the biblical image of God. So far I think we are still talking to open-minded scientists, monotheists, Trinitarians, pantheists, and various agnostics, although we have resisted contact with specific historical events and the epistemology of faith. Most of the religious differences come from the views of specific historical events and unshared acts of faith. I hope that by the time we have finished the book we will give all of these groups and other searchers for understanding reasons for discussion with each other. Note that during the first 12 billion years until the emergence of *Homo sapiens,* it was not meaningful to think of man in God's image, or of a transcendent God interacting with man. With the emergence of humans and the human mind, there has been a radical change in the universe, locally at least, resulting in having someone to think about it. Everything changed with the emergence of the human mind and human culture; even God changed as transcendence emerged. That startling idea, which some will find heretical, is a main theme of this book.

## Chapter 2—Readings

Bergson, Henri, 1911, *Creative Evolution,* H. Holt.
Darwin, Charles, 1859, *On the Origin of Species.*
Elsasser, Walter, 1966, *Atom and Organism,* Princeton University Press.
Holland, John H., 1995, *Hidden Order: How Adaptation Builds Complexity,* Addison-Wesley.
Holland, John H., 1998, *Emergence: From Chaos to Order,* Addison-Wesley.
John Paul II, 1992, *In the Emergence of Complexity in Mathematics, Physics, Chemistry, and Biology,* Pontifical Academy of Sciences.
Kant, Immanuel, 1985, *Critique of Pure Reason,* St. Martin's Press.
Morgan, C. Lloyd, 1923, *Emergent Evolution,* H. Holt.
Quastler, Henry, 1964, *The Emergence of Biological Organization,* Yale University Press.
Smolin, Lee, 1997, *The Life of the Cosmos,* Oxford University Press.
Teilhard de Chardin, Pierre, 1959, *The Phenomenon of Man,* Harper.
Weinberg, Steven, 1977, *The First Three Minutes,* Basic Books.

# 3

# The Twenty-Eight Steps

In the ground-breaking book on emergence *(Emergence: From Chaos to Order,* by John Holland) referred to in the last chapter, the author notes, "Despite its ubiquity and importance, emergence is an enigmatic, recondite topic, more wondered at than analyzed. What understanding we do is mostly through a catalog of instances." I accept Holland's subsequent caveat that "It is unlikely that a topic as complicated as emergence will submit weakly to a concise definition." Having said that, I shall indicate how we can discuss emergence, go a bit beyond Holland's generalization, and present a catalog of 28 observed instances that have emergence in common but vary over an enormous range in the agents, interactions, hierarchical levels, and character of the interaction rules and the pruning rules. The examples should illuminate the many meanings of emergence and, at the same time, show how this approach has commonalities and aids our understanding of how the world, as we now experience it, came to be.

The 28 examples of emergence that have been chosen are not completely arbitrary. They are selected to form an almost linear chronological sequence from the beginning of the universe we now occupy toward a conscious grasping for the future, a search for spirit, or something in that domain. The emergences outline a history of the cosmos from an anthropocentric point of view, where continuity is punctuated by emergences and each emergence gives some foresight into what may follow. We can then think about the special character of this sequence and raise questions, based on the selection rules, about chance or purpose in the unfolding universe. Thus our search for understanding emergence has the second agenda of asking how such a comprehension can illuminate those philosophical, theological, and personal yearnings for understanding that lie deep in the essential self of every man and woman. Is there a divine plan in the playing out of the history of this universe, or is history simply a series of very improbable, chance events devoid of meaning? Does the truth lie somewhere in between?

Our plan is threefold. In this chapter we will present an overview of the 28 steps that lead from the immanence of the laws of physics to the transcendence in human interactions and perhaps beyond. Once this scheme is outlined, we then present a more detailed analysis of each of the emergences and finally analyze the concept itself in terms of the instances. With this in mind, we will try to deepen our general understanding of emergence per se and revisit the recent examples at the top of the hierarchy: cognition, reflective thought, philosophy, religion, and spirituality. We end with thoughts of the dialog that the construct of emergence opens between science and religion.

There is some arbitrariness to the number 28. It is a compromise between viewing the major changes and including enough detail to reify the concept of emergence. Biologists are somehow usually splitters or lumpers. I have tried to steer a middle course between the extremes.

## Step 1. The primordium

This is the beginning, the emergence of something from we know not what. Here is the ultimate mystery of mysteries. Origins are clearly central to science, philosophy, and religion. The three principal beginnings are origin of the universe, origin of life, and origin of mind. In this step we will discuss the Big Bang and the first three minutes, as reconstructed from physics. The central idea is that our universe had a beginning, and rather remarkably we can attempt to reconstruct its history since that time, although we may never know "the thing in itself." Cosmologists and particle physicists have made great progress in this first emergence. Although we must divide it into domains, we will discuss the entire process in one emergence.

## Step 2. Large-scale structure

As the hot, dense, inflationary universe expanded and particulate matter appeared, density fluctuations in the number of particles emerged, which are the apparent cause of large-scale cosmological organization. This structure is now being mapped in the galaxies and even larger aggregates such as galactic clusters. This mapping is an active field of research in modern astronomy and astrophysics. Again, the reason for this organization is deep and enigmatic; however, we do know that the emergence of structure in the cosmos depends on ill-understood density fluctuations and the universal force of gravity.

## Step 3. Stars

The collection of primordial matter into stars begins as part of the large-scale structuring, in the process of gathering the hydrogen and helium into larger cosmic entities. Two operations are of primary importance: gravity, which transforms density variations into objects such as stars, and nucleosynthesis (or fusion reactions), which operates at the core of stars and in stellar explosions and converts the original light atomic nuclei into a collection of heavier nuclei in addition to electromagnetic radiation. There is a good understanding of the underlying stellar dynamics and nuclear physics, and there are observations on a large number of stars. The life and death of stars is thus surprisingly well understood. Stars emerge in the first instance from hydrogen and helium, and second- and third-generation stars add heavier elements that are cooked up in the interior of earlier stars and in the explosion of novae and supernovae.

## Step 4. The elements

When the soup of nuclei formed from more elementary particles, electrons, and photons, which is referred to as a dense plasma, cools sufficiently, the particles come together and matter as we know it emerges. A strong selection principle operating at this level, the Pauli exclusion principle, leads to an arrangement of electrons and nuclei that result in the periodic table of the elements, chemical bonding, and the different states of matter. The subdomain of physics that we designate as chemistry and physical chemistry thus appears or emerges at this stage. The primary entities for all subsequent steps are nuclei and electrons. The universe as a whole is characterized by a continuing and dynamic birth and death of generations of stars with an ever-changing elemental composition due largely to fusion reactions and radioactive decays. The pruning rule that generates all of chemistry is the exclusion principle, an extradynamical law of physics, which is the dominant factor in organizing the material world at temperatures below $3000°$ K. This principle is extremely important in all subsequent emergences and may hint at something deeper and provide clues to organization of higher levels.

## Step 5. Solar systems

After stellar condensations and explosions, second- and higher-generation stars emerge containing more and more heavy elements, and new factors such as stardust and various force fields become part of the dynamics. These forces, plus the ever-present gravity, plus conservation of angular momentum, lead to particulate rings and asteroids and planets and other structures condensing out of space debris, and orbiting the condensing stars. Solar systems emerge. These are novel organizations of matter in the cosmos; they have emerged with the formation of second- and third-generation stars and could not have occurred before that time.

## Step 6. Planets

Among the interstellar materials that orbit stars, some components condense into large satellites that are designated planets. The condensation processes and subsequent processing lead to geophysical structures, often spherically symmetrical and composed of shells. For earthlike planets, gravity, centrifugal separation, melting, and various processes involving the macroscopic properties of the collection of elements out of which the planet is made are operative. These include phase changes and crystallization. In the case of Earth, the solid core, molten outer core, mantle, and the other layers emerge. Dynamic processes are driven primarily by various heat sources such as nuclear decay and gravitational potential energy. Large-scale planetary shell structure emerges. There seems to be something quite special about Earth.

## Step 7. The geospheres

All the processes that took place in the later accretion and weathering and outgassing of the prebiotic planet led in aggregate to the emergence of the lithosphere (rocks), hydrosphere (water), and atmosphere (gas). The operative processes are mechanical, thermal, hydrodynamical, chemical, aerodynamical, and meteorological. A distinction must be made between the geospheres before life and after the origin of life, when a new, catalytically active geosphere, the biosphere, altered the relations among the other geospheres. In the formation of the planet and shortly thereafter these major geological structures appeared.

## Step 8. The biosphere

In order for cellular life to arise, the right collection of chemicals must be captured in physical space and informational space. Either capture might come first. Physical separation might be accomplished either by adsorption on the surface, by trapping in porous structures, or by capture in colloidal vesicles. Informational trapping is accomplished by the network of chemical reactions run by available energy and catalysts. The combination of network chemistry and spatial chemistry leads to the earliest protocells, entities capable of replication. The rules are the principles of organic and physical chemistry. The pruning principles are the logic of autocatalysis and self-replication. I suspect a deeper logic in the development of metabolism. Self-replicating protocells emerge, and with them competition for resources appears and the world becomes Darwinian.

## Step 9. The prokaryotes

The universality of so many features of intermediary metabolism and macromolecular synthesis suggests that all current life is descended from a common universal ancestor, and the study of the ubiquitous biochemical features of all present life can take us back to that organism. With the arrival of the first replicating protocells, life is clearly, as noted, in the competitive Darwinian domain, where the dominant selection principle is fitness as defined by replication and survival. Returning to the protocells, a great degree of macromolecular organization occurred, giving rise to the prokaryotes. This emergence probably occurred quite rapidly on a geological time scale. There was something very special about this step, because prokaryotes were the sole biota for about two billion years, and they are still the dominant biota. From examining the present version of the phylogenetic tree, the first emergent form was probably a thermophilic autotrophic bacterium. An enormously robust and subtle collection of genetic molecular apparatus emerged that is central to the character of all succeeding life. This basic molecular biology has been fixed for about four billion years.

## Step 10. Cells with organelles: eukaryotes

Two billion years of evolution led to a wide variety of prokaryotes that occupied all available niches. Two subclasses evolved with major structural chemical differences: these are the eubacteria and the archebacteria. Although the eubacteria probably came first, there is some uncertainty. The two taxa of bacteria vary in a number of subtle chemical features. Something then happened that allowed prokaryotic cells to combine by a process of one cell being engulfed by and living within another, called endosymbiosis. This led to: cells with membrane-bounded organelles, the origin of complex chromosomes, the process of meiosis, and the accompanying massive exchange of genetic material. More complex cells and sex emerged. This was a major change in cell organization, and is still reflected in the existence of two great groups of organisms distinguished by cell organization: prokaryotes and eukaryotes. Endosymbiosis must have followed a change occurring in cell chemistry, weakening the barrier of cell walls and making cells much more open to exchange of large-sized components. This may have made eukaryotes possible. Selection rules involve the fitness of the combined entities, and the emergent cells are known as protists or protoctists (Margulis, 1998). Endosymbiosis with previously unknown combinations of parts is a new way of creating novelty.

## Step 11. Multicellularity

With the facile exchange of material between cells that was possible for eukaryotes, as contrasted to prokaryotes, it became advantageous for cells to aggregate and to specialize. Multicellular organisms eventually emerged with morphogenetic programs that directed single-celled fertilized eggs to develop into structured multicellular collections with properties appropriate to their respective niches. There are three general kinds of multicelled organisms: plants, animals, and fungi. Fitness for each involved modes of obtaining food. Differentiation among the cells into various types is one of the emergent properties. Whole-organism form or morphology enters the world as a hierarchical step up from cellular morphology. Signaling molecules and surface recognition seem key features in directing this differentiation.

## Step 12. Neurons

At this point we focus on the animals, multicellular clusters of specialized cells that in aggregate get their food by eating other organisms. The cell-to-cell communication of information in the simple animals and other multicellular collections is largely chemical. Signaling molecules secreted by one cell can diffuse into neighboring cells, carrying messages. Diffusion is a rather slow process, especially at distances of several cell diameters, so diffusion-limited animals were of necessity either small or very slow, which is not optimal for being either predators or prey. Over a period of a billion years or so, animals evolved a new kind of cell to allow the rapid transmission of information over long distances. The cell consists of two informational chemical processing units connected by a long electrical transmitting component, the axon. The chemical events at one end trigger a rapidly transmitted electrical signal that causes the release of signaling chemicals at the other end. Thus the various parts of an animal can communicate with each other rapidly while keeping the historically deeply embedded chemical mode of information transfer. The chemical-electrical-chemical unit is a cell called a neuron. This cell type probably evolved from an epithelial cell. Macroscopic animals capable of growing much larger in size then emerged, able to respond rapidly in getting food or fleeing from predators. The selection rules were in the domain of fitness. The major emergent property is neuronal information transfer through neural nets.

## Step 13. The emergence of two subkingdoms of animals

Among the early animals with neurons were cylindrically symmetrical coelenterates such as jellyfish and worms. Some of the latter were bilaterally symmetrical and possessed a quite elaborate nervous system. These early animals developed a pattern of morphogenesis, embryological development, that has persisted among members of this kingdom. Fertilized eggs undergo repeated cell divisions and form a cluster of cells, first a spherical blastula, then a gastrula that elongates into the axial form characteristic of the worms. During the morphogenetic development of the primitive embryonic gut, an opening forms that becomes the mouth. Animals with a single gut opening are called protostomia. This group includes most of the invertebrates. In the development of a second group, fusion of the middle part of the elongated blastopore creates a mouth and a separate anus in many adult forms. They are called deuterostomia. The common ancestor of both groups is probably, but not certainly, a flatworm or platyhelminth. It is one of those great evolutionary branchings that emerge from primitive animals and results in two major subgroups. We shall focus in subsequent emergences on the deuterostomia, a highly diverse collection including us and our hardly obvious relatives Echinodermata (starfish, sea urchins), Chaetognatha (arrow worms), Brachiopods (lampshells), and Chordata (vertebrates like us and more primitive chordates). The validity of the grouping is clear in the observed data both embryological and molecular, but the selection principle that led to these two organismic forms is not at all obvious. In this case, we may find the emergence of the deuterostomia difficult to explain, but chordateness is a property of great importance in the subsequent evolution leading to the brain and related features. The fossil record of these soft organisms is very incomplete, and the evolutionary tree of the early Cambrian Age, in which a number of these taxa emerged some 550 million years ago, is quite speculative.

## Step 14. Chordates to vertebrates

Chordates are generally characterized by a rod-shaped notochord, a hollow neural tube, gill slits at some stage of development, a segmented body, bilateral symmetry, and a closed blood system with a pump, the heart. One of the evolutionary radiations of the chordates is the Vertebrata, which replace the notochord by a cartilaginous or bony vertebrate (the backbone) and a series of body transformations related to segmentation and cephalization. Vertebrates generally have a head and tail end, and there is a selection operative to move sensory function, control function, and ingestion forward along the vertebral axis toward the head. In the neural system the cephalized brain emerges, and this is the core of many further emergences. Indeed, one of the features of subsequent evolution is a progressive enlargement and complexification of the brain. Cephalization and complexification are generalizations of vertebrate evolution. What selection criteria lead to this feature is not certain. In any case, the first animals with brains emerge at Step 14. Brains have also evolved in invertebrates: lobsters and other arthropods, cephalopods such as octopus and squid, and other taxa. The present-day descendants of the most primitive of chordates, the amphioxus, are the lampreys and hagfish. An emergent property related to cephalization is cognition or consciousness that is the animal mind. Some biologists believe it appears at an even more primitive level. Indeed some would trace awareness back to the preneuronal forms or even to prototistan.

## Step 15. Fish

There follows a long evolutionary history from the jawless fish to the cartilaginous fish and the bony fish. From some of those fish living in shallow waters and developed lungs, life on land became a possibility. This depended on the geological occurrence of continents and islands, and the ecological development of terrestrial plants and invertebrates as a food source, and the evolutionary development of appropriate physiological structures. During the evolutionary radiation of fish, highly developed vertebrates emerge. The vertebrate body plan emerged in the evolution of fish.

### Step 16. Amphibians

The transition to land involves the appearance of animals with aquatic zygotes and larval phases and land-dwelling adults. Around 390 million years ago, such forms emerged that are called amphibians because of their living in two geospheres. This is clearly an evolutionary emergence, but one driven by geological changes and the appearance of new ecological niches. From the first fish to the first amphibian took about 80 million years. Frogs and salamanders still represent this transitional group of animals.

### Step 17. Reptiles

The next major emergence was driven by the need to overcome the constant threat of desiccation or drying out that faces amphibians at most stages of the life cycle, particularly early in development. For the embryonic stages, this problem was solved by the development of the shelled egg with its own food storage in the form of yolk. This mode of morphogenesis required internal fertilization so that the zygote and yolk were present before the shell was laid down. The second answer to drying out was a keratose scale covering the skin and serving as a barrier to water. Under this kind of selection, sometime around 280 million years ago, a new taxon emerged: the reptiles. Note that this emergence from amphibians took about 100 million years. Major evolutionary changes often require many generations, especially when several alterations in anatomy and physiology are required.

### Step 18. Mammals

While it seems clear that both birds and mammals are evolutionary offshoots of reptiles, one cannot trace a simple single path from reptiles to mammals, and a number of fossils have been referred to as mammal-like reptiles. The reptiles were clearly dominant until 65 million years ago, when there also existed small, probably carnivorous forms that are the ancestors of most modern mammalian forms. They had been around for a long time. The situation is complicated by the presence of a variety of mammalian types: egg-laying monotremes, the pouch-developing marsupials, and the placentals. In any case, the mammals have emerged as the major terrestrial vertebrates over the past 65 million years.

## Step 19. Arboreal mammals

In the last few steps, the idea of a niche has developed as an ecologically based component of selection. A niche is a place to live and a way to survive with respect to food, water, protection, and any other needs of organisms. A rule of evolutionary biology is that, given a niche, members of a taxon will evolve to compete for that niche. So that, given a forest, which emerges from the evolution of trees and plants, arboreal animals including mammals will evolve to occupy that niche. They could be climbing or flying mammals, finding food in the niche as fruit eaters, leaf eaters, insectivores, or predators of other arboreal animals. Living in the trees places certain requirements on a species, such as location of the eyes for stereoscopic vision, grasping hands, feet, and tail, and use of the tail for balance. And so, from the existence of forests and the presence of stem mammals, the arboreal mammals emerged: they are insectivores, bats, and prosimians. The selection principle was survival in the forest ecosystem.

## Step 20. Primates

The emergence of the primates from some arboreal prosimian ancestors was the next step. Primates appear to be the result of several selections such as enlargement of the cerebral cortex and its behavioral consequences, including a social organization that extends maternal care and training of the young. The evolutionary radiation led to the emergence of a wide variety of simians, including both the Old World and New World monkeys.

## Step 21. The great apes

At some point, the drying of the African forests and their conversion to grassland exerted extreme pressure on the arboreal primates. Competition in the shrinking forest was fierce. The cause of the drying was apparently geological and meteorological, but the consequences were ecological and severe in terms of competition. The changes along one evolutionary pathway included adaptation to life on the savannah, increase in size, loss of tail, change toward terrestrial as contrasted with arboreal modes of locomotion, increase in brain size, and development of modes of social organization. The apes emerged.

## Step 22. Hominids

On the grassland the number and variety of niches were much smaller as compared with the forest. Competition in the more uniform environments became even more intense. The extended period of infancy resulted in increased divergence of sexual roles. Social organization led to improvement in cooperative methods of acquiring food. Fully bipedal locomotion proved to be advantageous, and the increase of cognitive skills with increased brain size added to fitness. The hominids, probably beginning with the African *Australopithecus* or some earlier form, emerged from the apes. Evolution under these circumstances moved at some stage from the purely physical to take on major cognitive and social components. At this emergence the evolutionary rules appear to have changed substantially, with major learned components entering the picture.

## Step 23. Toolmakers

Steps 10 through 20 or 21 define the classical domain of evolutionary biology. Emergences are dominated by the interplay of organic forms and occupancy of niches. All of biology is thus interactive because a niche is both biological and geological, and it is often difficult to distinguish the components. A change in emphasis starts with hominids perhaps even somewhat earlier: the organisms are able to alter their environments and change habitats in conscious ways. The construction of dams by beavers is an example of this behavior. This feature eventually shows up in making tools, effective instruments for altering the environment. The first emergent tool-making hominid species that we have evidence of is *Homo habilis*.

## Step 24. Language

The transition from *Homo habilis* to *Homo sapiens* probably occupied about two million years, as did the prior transition from *Australopithecus* to *Homo habilis*. Knowledge of what occurred in this four-million-year period is at the core of the questions of who we are and where we have come from, but it has been a recognizably difficult field of research because of the scarcity of fossils. It was once improperly called the problem of the missing link. The evidence on intermediate forms is thus scanty and sometimes conflicting. The emergence of language is clearly the part of the problem of greatest subtlety. There are no fossil remains of language; indeed, there are no fossils of organs of speech. In many ways language is key to subsequent development. Without language, effective cultural transmission of information would not be possible. Darwinian evolution argues against the transmission of acquired characteristics. With language, a kind of informational transfer between the generations or social Lamarckianism in the form of learning became incorporated into all of culture and allows for the subsequent steps. From the group behavior of hominids, language emerges at an uncertain time.

### Step 25. Agriculture

It appears likely that for a long period, perhaps a quarter of a million years, *Homo sapiens* were hunter-gatherers, and only about 10,000 years ago did agriculture emerge in the form of domestication of food animals and the cultivation of food plants. An agricultural society could live in a relatively fixed locale, although nomadic herders may be intermediate between hunter-gatherers and agrarian society. While agriculture is clearly biological in terms of food plants and animals, it is an example of humans' controlling their own niche. Agriculture is an information-laden emergence. It radically alters the habitat.

### Step 26. Technology and urbanization

It is not possible to separate all the steps from the sod to the spirit in a discrete manner, but accompanying the expansion of agriculture was a developing technology that included the science of mechanics: the lever, the wheel, and the inclined plane. This required a certain amount of applied mathematics. The building of fixed structures led in turn to urbanization. Weapons were also improved, as the result of the increasing competition between human groups. By 5,000 years ago, enough technology had developed to lead to large stone buildings: urbanization emerged. There are those who would argue that some cultural aspects of the human spirit carried by the oral tradition predate agriculture; it goes far back into our hominid past. Again, it is not possible to maintain a strict chronology. Nevertheless, the emergence of technology is a major step between hunter-gatherers and modern civilization.

## Step 27. Philosophy

About 5,000 or more years ago a sufficient understanding of the laws of nature had been developed and transmitted so that some individuals undertook to search for an understanding of who we were and how we arrived at where we found ourselves. There was reflexive turning back on some of the earlier emergences in an effort at self-understanding. The ongoing search for fundamentals in all domains that we designate as philosophy and natural philosophy emerged. This human activity includes science, religion, myth, and all those approaches that enter into the broadest conception of understanding the world. This mode of thought, divorced from the practical and representing a real novelty, seems to have sprung forth all over Earth and in almost every culture.

## Step 28. The spiritual

It is unreasonable to assume that all emergences have ended. Teilhard de Chardin has provided the key to the present emergence of mind by suggesting the evolution of the spirit as the next evolving stage. Chapter 32 will ask what we mean by the emergence of the spirit. At the very least, it is an attempt to ascribe meaning to the other emergences. It is the intersection of what has come to be science and religion.

Having taken a quick view of the nature of emergence, and having scanned the series of emergences between the beginning and the present, we now proceed to a somewhat more detailed view. The purpose is to explore the emergences in a search for generalization and understanding to provide a new science able to utilize the tools presented to us by the computer age.

In the 28 examples being covered, the selection criteria appear sufficiently different that lumping them all as emergences may seem like too broad a concept of an emergence, yet there are features in common. Each of them alters the universe and leaves it different after its occurrence. Each was incompletely predictable before it happened, and each provides the foundation for the next emergence. Nonetheless, the first seven emergences, the next twelve emergences, and the last nine seem to group. They each have a somewhat different character.

This overlap between the concepts of emergence and evolution characterizes Steps 9 through 20. Evolution is the overall process, while emergence characterizes the punctuations. We often know the emergences, even if we are uncertain about the details of how they come about.

Steps 1 through 7 are in the domain of the physical sciences. Step 8 is transitional. Steps 9 through 20 are biological, and Step 21, the appearance of our primate ancestors, is transitional. Steps 22 through 26 are cultural, the emergence of societies of hominids. Step 27, reflective thought, is transitional, leading to Step 28, where the theme is the world understanding and governing itself.

This is not a new finding. Some 60 years ago, the great student of philosophy Ernst Cassirer wrote a very profound study called *The Problem of Knowledge: Philosophy, Science, and History Since Hegel*. The three parts are called "Exact Science," "The Ideal of Knowledge and Its Transformations in Biology," and "Fundamental Forms and Tendencies of Historical Knowledge." For us terrestrials, this corresponds to the first eight billion years, the next four billion years, and the last 200,000 years. Our domains of knowledge and our classification of emergences correspond to Cassirer's categories. The major origins are origin of the universe, origin of life, and origin of mind. Within each of the superdomains are families of emergences. The tools of understanding the domains are physics, biology, and history.

Again we see a continuity in our global ways of understanding. While recognizing the newness of contemporary thought, it occurs within the framework of an intellectual tradition. I find this satisfying.


## Chapter 3—Readings

Cassirer, Ernst, 1950, *The Problem of Knowledge: Philosophy, Science, and History Since Hegel*, Yale University Press.
Margulis, Lynn, 1998, *Symbiotic Planet*, Basic Books.
Stebbins, G. Ledyard, 1982, *Darwin to DNA: Molecules to Humanity*, W. A. Freeman and Company.

# 4

# The First Emergence: The Primordium—Why Is There Something Rather Than Nothing?

For thousands of years each culture around the world has developed a creation scenario or series of scenarios that have developed into ontological and theological schemata explaining existence itself. While the encyclopaedia refers to these views as "myths and doctrines of creation," they play an important role in the self-image and religion of each society. We may conclude that speculation about origins is an essential part of the human condition. There are a number of these origins to be explained: matter and energy, celestial objects, and living beings. We add the origin of mind to other beginnings, for in the present context it is important to know when and how mind entered the description of the universe.

The dominant creation views in Western culture, which we regard as scientific rather than mythological, largely stem from astrophysics, particle physics, and cosmology. These sciences allow us to formulate a remarkably detailed picture of the first emergence that we can treat as a singular event but could equally well divide into a series of happenings.

In the late 1920s, astronomer Edwin Hubble undertook a study of the Doppler shift in light coming from a number of stars. The Doppler effect results in light from objects moving away from the observer shifting toward the red, and light from objects moving toward the observer shifting toward the blue. Since atomic line spectra should be the same everywhere in the universe, we can measure the velocity of each star with respect to the observer along a line from the Earth to the star. Hubble concluded that all galaxies were moving apart from each other at a speed proportional to the distance between them.

By the 1950s we had enough data to conclude that the entire universe is rapidly expanding with all objects flying apart at high rates of speed. If the universe is everywhere flying apart at ever-increasing speeds, then one should be able to follow back the trajectories in time to the point of convergence, to the time when it all started. This led to the idea of the Big Bang: the cataclysmic event some 12 billion years ago when it all began as a giant explosion of totally unimaginable proportions.

The earliest events have been designated the inflationary universe. Cosmologists are able to present a consistent scenario of the inflation.

For the first $10^{-43}$ seconds, the stuff of the universe was hot and so dense that the distinctions of matter and energy disappeared and all of the four kinds of forces merged into a single, unified force. The four forces are: gravity operating at a large scale, the strong and weak forces that govern nuclear stability and decay, and the electromagnetic force that govern most of our ordinary world including chemistry and biology. At $10^{-43}$ seconds the force of gravity split off from the other forces. At $10^{-35}$ seconds, the strong force froze out, separating it from the electromagnetic and electroweak forces that remained unified to one ten-billionth of a second when it split into electromagnetism and weak force. The transition at $10^{-35}$ seconds was accompanied by an enormous expansion of volume by a factor of some $10^{100}$ or more. This period has been designated the inflationary universe.

The numbers and descriptions we have just given are beyond my comprehension and ability to visualize, but they represent the cosmologists' best view of the earliest universe. We cannot as scientists discern the reason for the step presented above. We can, however, reconstruct the events with surprising conviction. Parenthetically we might ask whether it is possible to go back to theorize about a time before the Big Bang. A few years ago I would have thought not, but astrophysicist Lee Smolin suggested in *The Life of the Cosmos* that we might be able to develop a theory that includes the origin of universes. I now realize that no questions lie beyond the imaginative probings of scientists. But for present purposes we will confine ourselves to this universe and start with the first emergence, the appearance of forces and matter and energy from a state we know not.

The universe, we believe, began in a very hot, very dense state some 12 billion years ago. There is little to be said of the instant of creation, because the temperature and density approached infinity, where our theory breaks down. The very idea of an instant of creation, nevertheless, is important to some theologians as a point of contact with science. Discussions of creation and eternal existence held an important place in medieval philosophy and theology, and to some they still do. There are astrophysicists who believe in a steady-state universe. The beginning is still a mystery, but concerning the inflation at short times after the beginning, there are the results of calculations. The assumptions were made that theories of thermodynamics and particle physics can be carried into almost any domain of temperature and density. An underlying assumption is that the laws of physics hold everywhere in the universe and for all time.

Following inflation the universe was a very dense soup of quarks and electrons. At about $\frac{1}{100}$ of a second, the temperature dropped to 100 billion degrees Kelvin, and the density decreased to 3.8 billion times that of ordinary water. The quarks condensed into hadrons, particles like protons and neutrons that are subject to the strong force. There were also leptons such as electrons and neutrinos.

By $\frac{1}{10}$ second, the temperature dropped to 30 billion degrees, and the contents were dominated by electrons, positrons, neutrinos, antineutrinos, and a much smaller number of protons and neutrons. By three minutes, light nuclei were stable and helium and lithium nuclei joined the hydrogen nuclei. By 34 minutes, the temperature was 300 million degrees, and the universe consisted mainly of neutrinos, antineutrinos, electrons, protons, and helium nuclei. About 20-30 percent of the heavier particles are helium, calculated by weight. There was also a small amount of deuterium.

The very hot, dense state continued to expand and cool. At certain temperatures, various phase changes occurred that altered the nature of the particles and the coupling forces between them. At temperatures of 3000° K and lower, electrons and nuclei formed stable atoms. This removed free electrons and effectively uncoupled the major interaction between electrons and photons, since their interaction is enormously decreased when the electrons are bound. The photons continued to expand as black-body radiation at 3000° K, which cooled by Doppler shift to the present 3° K black-body background radiation that fills the universe. By a study of this radiation, we conclude that the uncoupling occurred about 700,000 years after the Big Bang. This is a point that can be established from present-day measurement. It turns out that black-body radiation and the universe's abundance of hydrogen and helium are contemporary measurements used to check the theory of the original expansion. We can also calculate that at the 3000° K transition, the universe was 1,000 times smaller than at present.

The theories on which we base these views will doubtless change with time, and understanding the universe from a scientific point of view is always changing. For example, in 1998, new data from very distant galaxies suggested that the universe is expanding at a greater rate than previously believed, requiring the introduction of a new repulsive energy to explain the data. This may alter our views of the age of the universe. While new developments will change our detailed understanding, I believe that the basic events of the 28 steps we are discussing will stay intact and provide a framework within which to search for other origins, including the spirit.

Steven Weinberg's book *The First Three Minutes,* a physicist's view of the first origin, ends on a dispiriting note. He writes:

> It is almost irresistible for humans to believe that we have some special relation to the universe, that human life is not just a more-or-less farcical outcome of a chain of accidents reaching back to the first three minutes, but that we were somehow built in from the beginning. . . . It is very hard to realize that this all is just a tiny part of an overwhelmingly hostile universe. It is even harder to realize that this present universe has evolved from an unspeakably unfamiliar early condition, and faces a future extinction of endless cold or intolerable heat. The more the universe seems comprehensible, the more it also seems pointless.

Weinberg's pessimism, I believe, comes from examining only one of the 28 steps. I think that his angst is premature. We should look at all the emergences, even emergence itself, before assessing meaning or lack of meaning in the evolving universe. To some, the search for meaning is our human imperative.

In any case, the beginning has a reality all its own. The beginning state is one of great simplicity. Yet lurking within it must be all the forthcoming complexities; our task is to think about how they play out.

Over the history of Western civilization, our views of the Creator and creation have been anthropocentric and shrouded in human metaphors. The nature of the creative event of science is so cosmic and so cataclysmic that we are forced to reexamine what we can mean by the Creator. Small changes won't do. Here science is calling for a major paradigm shift in our conceptualization of God the Creator. It is a challenge that theologians cannot avoid. The God of the Big Bang is almost impossible to reconcile with the God who rested on the seventh day.

Science and Western religion agree on one thing: our universe had a beginning. Science also suggests that it may have an end. So do some religious doctrines. I find it difficult to speculate on what might happen millions or billions of years in the future. The epic of human civilization is only about 10,000 years old, and we have many miles to go before we sleep.

## Chapter 4—Readings

Guth, Alan H., 1997, *The Inflationary Universe,* Addison-Wesley.
Trefil, James, and Hazen, Robert M., 2000, *The Sciences; An Integrated Approach,* 3rd ed., John Wiley and Sons.
Weinberg, Steven, 1997, *The First Three Minutes,* Basic Books.

5

# The Second Step: Making a Nonuniform Universe

The primordial event led to a uniform sea of particles and photons flying apart, with the whole system heading for a low-density heat-death of the universe as was predicted by Rudolph Clausius's version of the second law of thermodynamics formulated in the mid-1800s. But as we look around, we see that a uniform expansion is clearly not what happened. For example, we are here. We gaze on the heavens today by eye and scan the sky by instruments and see innumerable stars, galaxies, galactic clusters, quasars, pulsars, and other celestial entities. How did the uniform sea of submicroscopic particles fleeing from one another give rise to the large-scale structure of our present universe and ultimately to our earthly home?

The one known force that counters the flying apart of the universe is the universal attractive pull of gravity that operates between all matter. Even the postulated "dark matter," which we cannot see because it doesn't interact with photons, is supposed to be felt by its gravitational effect. Although Newton's introduction of this enigmatic attractive force of gravity ushered in the beginning of our understanding of celestial mechanics, the motion of planets, gravity still remains at best incompletely understood, perhaps the least understood of all the physical forces that have been studied. In a completely uniform universe, gravity would simply slow down the expansion. Indeed, if there were enough matter in the entire universe, gravity would at some time reverse the flying apart and condense all the substance of the cosmos back into a dense, hot mass, a condition sometimes designated as the Big Crunch. All of this could possibly happen without any structure at all appearing. The fact that structure has occurred must be deeply embedded in the laws of nature or the laws of emergence.

In order for gravity to give rise to large-scale structure, another factor must enter: there must be density fluctuations in the expanding universe. At any point in an absolutely homogeneous sea of matter and energy, the gravitational force in any direction would be canceled out by an equal gravitational force in the opposite direction. This will be true in all directions, and the density will remain uniform in space, even though it decreases with time during the expansion. If, however, a density fluctuation of macroscopic dimensions occurs or appears on expansion, then the region of high density will be an attractive center, and the surrounding material will be pulled toward it. This will make the core a zone of even higher density, and the condensation will continue until some countering forces appear. Depending on the scale of density nonuniformity, the resulting stars, galaxies, and the rest of the large-scale structures that fill the universe will emerge.

When we scan the sky, there is, as noted above, a vast array of structures that have arisen from inhomogeneities or fluctuations in density. Here we face an enigma. There is no universally agreed-on explanation of the cause of the variation in density; there appear to be several possibilities. The theory by which we explain the first three minutes involves, at some point, equilibrium states that one would expect should be spatially homogeneous. The $3°$ K background radiation that fills the universe appears to be uniform in all directions, suggesting, but not necessarily requiring, homogeneity at 700,000 years after the Big Bang. (Some small nonuniformities are now under study, so this view may change.)

Among cosmologists there appear to be three related but not totally consistent theories of density fluctuation. One postulates that quantum fluctuations appeared very early in the process and somehow persisted. A second theory speaks of a phase change in the properties of space on inflation that broke the uniformities, and the third refers to gravity instabilities. While I am unable to critique these views of astrophysicists and cosmologists (indeed, I am not sure that I fully understand them), it is clear that a variation in density occurred, and the emergent property is the large-scale structure of the universe.

At present, groups of astronomers are busily engaged in mapping the heavens to catalog the large-scale structure of the cosmos in the hope that this will provide the experimental background to lead one day to a satisfactory theory of significant density fluctuations. Indeed, this is a major activity in contemporary astronomy. For now, we are presented with yet another mystery.

Observation of structure demands that there must have been inhomogeneities, and theory leads to a picture of a uniform expansion. For the moment, we are looking for new developments in theory. We are sure that large-scale structure emerged, since it is there, and we don't know why. This is a most unsatisfactory situation of the kind that drives the creativity of some scientists.

In steps one and two of our catalog of emergences, we are dealing with events that took place 12 or so billion years ago. With a great many missing bits of information, we try to reconstruct what happened when the universe was born. Birth itself seems like a strange metaphor. In what follows, we will try to reconstruct the subsequent steps that led to creatures like us who are able to look back across the eons and ask these questions about the first moments. I find it curious that there seems to be a relation between the beginning and our trying to understand the beginning. The Big Bang makes creationists of us all, although not necessarily of the fundamentalist sort. This is a difference between those who "know" the answer and those who are passionately seeking the answer.

For theists, deists, atheists, pantheists, and agnostics alike, the events at the dawn of time are truly awe-inspiring. Whether one wants to formulate a grand unified theory of everything or get down on one's knees to pray, the beginning cannot be ignored. The Big Bang and the density fluctuations are shrouded in mystery and may remain so shrouded forever. I accept these uncertainties and go on to the next stage. It is not necessary to live with the hubris of knowing everything.

I would argue, however, that the way that we ourselves have emerged along with all the other emergences calls upon us to examine the emergences and seek explanations. If I call this act trying to know God's mind, I am using a metaphor that resonates with the last few thousand years of human history. It also asks me to go beyond the obvious. It is a viewpoint that converts science from a profession to a vocation, a calling, and I find that quite satisfying. It certainly enriches my conversations with my theological friends.

Thoughts along these lines were recently expressed by Pope John Paul II in the 1998 Papal Encyclical, *Fides et Ratio* (Faith and Reason), in the following words:

> Finally, I cannot fail to address a word to *scientists,* whose research offers an ever greater knowledge of the universe as a whole and of the incredibly rich array of its component parts, animate and inanimate, with their complex atomic and molecular structures. So far has science come, especially in this century, that its

achievements never cease to amaze us. In expressing my admiration and in offering encouragement to these brave pioneers of scientific research, to whom humanity owes so much of its current development, I would urge them to continue their efforts without ever abandoning the *sapiential* horizon within which scientific and technological achievements are wedded to the philosophical and ethical values which are the distinctive and indelible mark of the human person. Scientists are well aware that "the search for truth, even when it concerns a finite reality of the world or of man, is never-ending, but always points beyond to something higher than the immediate object of study, to the questions which give access to Mystery."

The Pope's views seem to have been strongly influenced by those of Pierre Teilhard de Chardin. Albert Einstein, has been quoted with a similar theme: "I want to know how God created the universe. I am not interested in this or that spectral line or other phenomenon. I want to know God's mind. The rest are details."

If there is a self-glorification in trying to know the mind of God, I would suggest that it is shared by scientists and theologians. Different epistemological approaches may be used, and these can form the subject matter of the dialog we are seeking.

### Chapter 5—Readings

Einstein, Albert, 1981, as quoted by F. S. C. Northrop in *Prolegomena to a 1985 Philosophiae Naturalis Principia Mathematica*, 1985, Ox Bow Press.
John Paul II, 1998, *Fides et Ratio,* Papal Encyclical.

# 6

# The Emergence of Stars

The condition of the universe after the uncoupling of photons and charged particles and the joining of electrons and nuclei can be described as a vast sea of hydrogen atoms, helium atoms, neutrinos, and photons, spreading out and developing (for presently ill-understood reasons) variations of density over a whole range of sizes. Within this space, filled with the simplest molecules, were molecular clouds held together by gravity because they were surrounded by zones of lower density. Under the influence of this universal and space-filling attractive force, the hydrogens and heliums moved closer together. Energy was conserved, and as the atoms got closer the gravitational potential energy of the system decreased. The potential is proportional to the reciprocal of the distance between particles. This decrease must have been balanced by an increase of kinetic energy to preserve the conservation of energy. As a consequence, the molecules moved faster. Temperature increased since it is a measure of the kinetic energy of the moving molecules. The condensing clouds of gas thus continually became ever hotter and denser with time. They were becoming protostars.

Any object hotter than its surrounding begins to glow, emitting what is known as black-body radiation. The higher the temperature of an object, the shorter the wavelength of the radiation maximum. This was experimentally determined in the late 1800s as Wien's law: $\lambda_{max} = \text{constant}/T$. Many years ago I heard a children's scientific teaching record with the refrain, "The color of the star, you may be sure, is mainly due to its temperature." And so it is. By determining the emission spectrum of a star we can determine its surface temperature.

As the thermal energy of the condensing mass diffuses out to the surface to radiate away, it is clear that the core must be hotter than the surface with a gradient of temperature between them. So the earliest protostar, or first-generation star, was a sphere mainly of condensing hydrogen and helium, hottest at the center and cooling down toward the surface where energy was being radiated to the rest of the universe. Were it not for another phenomenon that comes in during the process of star formation, the protostar would keep getting smaller and brighter.

The new phenomenon that comes into play is nucleosynthesis or fusion reactions. As the temperature of the core rises, the constituent atomic nuclei finally move with sufficient speed that on collision they undergo nuclear fusion reactions. Very high speeds are required for these reactions. In our experiments we get these speeds with particle accelerators. Remember that temperature is a measure of particle velocity. These are the types of processes that take place in hydrogen bombs. Fusion reactions are of the following general kinds:

$$4H^1 \rightarrow He^4 + \text{energy}$$
$$3He^4 \rightarrow C^{12} + \text{energy}$$
$$C^{12} + He^4 \rightarrow O^{16} + \text{energy}$$

These nuclear processes do two things: they release further energy making the core even hotter, and they produce new kinds of atomic nuclei that have not existed before. An emergent property of star formation is the appearance of a whole new array of elemental nuclei, changing the composition of the cosmos. Matter in its present form is not eternal; it emerged. Because there are nuclei of different kinds, matter is informational in addition to its other properties.

The second source of energy that enters the process along with gravitational potential is the fusion energy released by the nuclear reactions. This results from Einstein's relation $E = \Delta mc^2$, in which the fused nuclei of the newly formed elements weigh less than the components that fuse, and $\Delta m$ is the change of mass. Photons from the very hot stellar core now exert a radiation pressure outward that finally balances the inward pull of gravitational condensation; a star is born. The star achieves a steady state of a certain size and temperature and continues to glow in that state, without condensing, sometimes for billions of years, until it uses up its nuclear fuel.

In general, the larger the mass of the cloud that gives rise to a star, the higher the temperature of the star and the shorter the time that it stays in a radiation-steady state. We are here discussing first-generation stars made entirely from hydrogen and helium. As these stars produce other elements and sometimes explode, there is a space debris of heavier elements, and subsequent generations of stars, using this debris, will have more complex processes because of their chemical composition. The size of stars ranges from one-fifth the mass of the sun to 18 times the mass of the sun, and surface temperatures range from 3200° K to 33,000° K. Smaller protostars would never get hot enough to go nuclear, and larger masses would burn up too quickly to become stellar objects.

After the steady-state period, stars may change type due to internal processes involving the utilization of hydrogen, helium, carbon, or other elements, or they may explode catastrophically. The typical life of a star was whimsically described (Shklovskii and Sagan, 1966) as follows:

> It is common, in astronomy, to refer to certain broad categories of stars both by their relative sizes and by their colors. The astronomical zoo is replete with "supergiants," "giants," "dwarfs," and "sub-dwarfs," but no individuals of ordinary stature, and a simple statement of stellar evolution often sounds like an excursion into the world of the brothers Grimm. A typical star begins life auspiciously, as a bright yellow giant, and then metamorphoses, in early adolescence into a yellow dwarf. After spending most of its life in this state, the yellow dwarf rapidly expands into a luminous red giant, jumps the Hertzsprung gap, and decays violently into a hot white dwarf. It ends its life, cooling inexorably, as a degenerate black dwarf. Few readers will recall the original title of this moderately depressing life history, but many will find it vaguely familiar.

To understand the underlying causes of the varied careers of the stars, we must discuss further astronomical observations and their interpretations.

When second- and higher-generation stars form, the condensing cloud includes not only hydrogen and helium but also various gaseous and particulate debris from catastrophes of first-generation stars. Nucleosynthesis in these stars and various novae and explosive events in the cosmos produce other distributions of elements, and there is a constant build-up of heavier elements in the universe. The relative cosmic abundance of the elements, at present averaged from the composition of the outer layers of stars, stony meteorites, and gaseous nebulae, is shown below in Table 2 and is based on hydrogen having a value of one billion ($10^9$).

The elements shown and rarer elements are the entities from which the world is built. Notice that the key elements of life are among the most abundant elements of the universe: carbon, hydrogen, nitrogen, oxygen,

TABLE 2: COSMIC ABUNDANCE OF THE
ELEMENTS

| Element | Symbol | Abundance |
|---|---|---|
| Hydrogen | H | $10^9$ |
| Helium | He | $6.3 \times 10^7$ |
| Oxygen | O | 800,000 |
| Carbon | C | 500,000 |
| Nitrogen | N | 100,000 |
| Neon | Ne | 93,000 |
| Magnesium | Mg | 45,000 |
| Silicon | Si | 32,000 |
| Iron | Fe | 22,000 |
| Sulfur | S | 16,200 |
| Calcium | Ca | 2,300 |
| Aluminum | Al | 2,140 |
| Sodium | Na | 1,900 |
| Argon | Ar | 1,200 |
| Nickel | Ni | 500 |
| Chlorine | Cl | 420 |
| Phosphorus | P | 320 |
| Chromium | Cr | 166 |
| Fluorine | F | 80 |
| Potassium | K | 76 |
| Manganese | Mn | 75 |
| Titanium | Ti | 35 |

phosphorus, and sulfur. The ability to form planets like the Earth is also dependent on the abundance of iron and silicon.

Elements that are quite rare on a cosmic scale may nevertheless be biologically important. Molybdenum is required for nitrogen fixation, and cobalt is a necessity for vitamin B12. Plate tectonics, which may be required for continuous life, requires uranium, thorium, and potassium-40 to generate, by radioactive decay, the heat required for the convective currents that are at the basis of tectonics. Elemental composition shows up in many ways.

Thus a collection of elemental nuclei that are found in stars and the stardust or particulate matter of the universe has emerged from gravitationally driven processes and nucleosynthesis. The nuclear reactions that give rise to the elements conserve charge, so that for every positive nuclear charge there is an electron. Above 3000° K or so, these form an ionized plasma, but at lower temperatures the electrons and nuclei form atoms.

The material released by stars and sent into interstellar space presumably collects together into small particles or star dust. This happens at temperatures far below 3000° K, so that in discussing these particles we move from the domains of nuclear physics and plasma physics into chemistry, the science of how nuclei and electrons come together to form molecules, crystals, and metals. The existence of nuclei and electrons does not guarantee the existence of individual atoms, but only of a sea of nuclei and electrons, so that the total of positive and negative electrical charge adds up to zero.

In this temperature domain, the interaction of electrons and nuclei is governed by quantum mechanics. The controlling factor is that electrons interact with nuclei in certain quantum states designated as orbits. The interaction of an electron with a nucleus is characterized by four quantum numbers. The quantum mechanical solutions are the interaction rules. They yield probability distributions of the electron about the nucleus. At this point, a new and deep law of physics, the Pauli exclusion principle, enters asserting that no two electrons in an atom or molecule can have the same four quantum numbers. This will be further discussed below.

The periodic table of the elements relating the chemical properties of the atoms to each other, chemical bonding of atoms, and crystal behavior then emerges. In a word, all of chemistry proceeds from nuclei and electrons interacting by a rule set that is pruned by the exclusion principle. In this case the selection principle is a nondynamic rule that selects a certain small set of states of matter from the inconceivably vast array of possibilities. Entities now interact as chemicals, subject to the set of rules that govern all chemical behavior. In terms of our sequence, we are moving from particle physics to chemistry that has emerged with stellar evolution. The periodic table is a different kind of emergence because it involves a new physical principle not derivable from dynamics. It takes us back to our discussion of the different kinds of emergence. The Pauli principle, by defining all chemical interactions organizes all the subsequent emergences. It may be a clue to higher-order unfoldings of the cosmos.

So, as the result of the stellar emergences, the universe consisted of a great variety of stars, galaxies, occasional novas, pulsars, quasars, black holes, and an array of known and unknown entities generated by the laws of gravity, mechanics, thermodynamics, and nuclear synthesis. Along with this vastly complicated cosmic collection of entities, something else has emerged: the nuclei of the periodic table of elements. As a result, chemistry presents itself with all its sophistication and all the vast possibilities of making new structures.

Also, on a timescale ranging from nanoseconds to billions of years, the universe is dynamic, with objects coming into existence and exploding or decaying. When we look out into the cloudless sky, the objects appear to be timeless, but that is only on the short timescale of human events. From a cosmic perspective, the universe is in constant flux. The universe is 100 million times the age of the longest-lived individual, one million times the age of civilization, and 250,000 times the age of *Homo sapiens*. It is difficult to grasp the time ranges we must learn to deal with. Yet if we are to ask about meaning, we just have to realize that some things take a lot longer than others. When Shakespeare wrote, "I am as constant as the northern star," he was dealing with the lifespan of an individual. On a longer timescale, the northern star changes as the Earth's axis of rotation precesses like a spinning top.

It seems to me antireligious to lose patience, as some people do, with a God who took 12 billion years rather than six days to create a universe. Cosmic time is not our time.

## Chapter 6—Readings

Margenau, Henry, 1977, *The Nature of Physical Reality*, Ox Bow Press.
Pauling, Linus, 1940, *The Nature of the Chemical Bond*, Cornell University Press.
Shklovskii, I. S., and Sagan, Carl, 1966, *Intelligent Life in the Universe*, Holden Day.

# 7

# The Periodic Table

Along with stars at the large-scale level, we have seen that the workings
of the cosmos produce, as a result of several processes of nucleosynthesis,
a wide variety of atomic nuclei varying in atomic number and atomic
weight. As the nuclei emerge from the stellar cauldrons and cool down,
they reach a temperature at which they can combine with electrons to form
atoms.

If you enter almost any chemistry classroom or laboratory, you will find
on the wall a chart of the periodic table of the elements. In a neat array
it provides a basic pattern for the understanding of all chemistry. It is
perhaps the most fundamental and best-known icon of all science, the key
to biology as well as all chemistry.

Atomic nuclei are characterized by an atomic number, the charge of the
nucleus divided by the charge of the proton. This number determines the
number of electrons that must join each nucleus to form a neutral atom.
As we have noted, the interactions between nuclei and electrons are governed by the rules of quantum mechanics and the Pauli exclusion principle.
Understanding these features is one of the great triumphs of physics and
physical chemistry. The application of the new physics of the early twentieth century to ordinary chemistry was elegantly formulated early in the
development of the theory in *Introduction to Quantum Mechanics* by Pauling and Wilson and *The Nature of the Chemical Bond* by Pauling. These
books are classics of scientific exposition that sum up the great understanding of nature that came to maturity in the first 40 years of the twentieth
century. Linus Pauling has been one of the great expositors of science, as
well as one of the great chemical researchers.

On occasion I refer to books that are probably too technical to be of interest to many readers of this more philosophical treatise. I do this with books that have had a deep influence on my own thinking, and I want them available to those readers who have a special interest in a given topic. And so my views of chemistry have been formed by the writings of Linus Pauling. They are there if you wish to seek out the roots. I recommend them strongly.

But for the rules of quantum mechanics, the world of matter would be a sea of electrons and nuclei at the lowest energy level. As understood by the quantum mechanical rules of interaction, the electrons are associated with nuclei in discrete orbits of quantized energy values. If the electron distribution of quantum mechanics stood as the sole governing principle, the world would be a sea of nuclei and quantized low-level electrons, but the Pauli exclusion principle means there are discrete associations of nuclei and electrons, so that the universe at lower temperatures operates as a world of ordinary atoms leading to chemistry, structure, and all other rules we are familiar with in working with ordinary material. The sophistication of distributing electrons in energy levels comes from the Pauli principle.

Before looking at the principle in detail, let us restate the rules of chemistry in terms of emergence. In the study of the physical states of systems of atoms, the agents are nuclei and electrons and the rules are quantum mechanics and electricity and magnetism. The pruning relations that severely limit the eigen states (allowable atomic configurations) of matter are the solutions to the Schrödinger equation and the Pauli exclusion principle. The emergent behavior is the content of the science of chemistry: the periodic table of the elements, the rules of covalent bonding, ionic bonding, and metallic bonding, and the bulk properties of solids, liquids, and gases.

The simple statement of the exclusion principle is that no two electrons in an atom can have the same four quantum numbers. This leads to an understanding of the shell structure of atoms, the facts of chemical valence, the spectra of atoms and molecules, and the structure of crystals. This statement of exclusion follows from the more general and sophisticated mathematical rule that all functions representing states of two electrons must be of the antisymmetric variety. This is a nondynamical principle that governs how electrons interact with each other, yet it influences their dynamical behavior. It is a pruning rule deep within the laws of nature that only permits behavior of a certain symmetry character. It selects a set of states from all possible states.

Another feature of the exclusion principle is that it begins to illuminate how the whole may be different from the sum of the parts. For the exclusion principle has nothing to say about the behavior of an individual electron, yet it applies to a system of two or more electrons. The Pauli principle is a way of understanding why entities show in their togetherness laws of behavior different from the laws that govern them in isolation. Since the principle is nondynamical, it is as if the second electron knew what state the first electron was in: for a law of physics, exclusion has a curious and somewhat noetic character.

The previous argument is worth restating. The emergence of the periodic table has a special character. The pruning rule is apparently a deep principle of physics, but it is unrelated to the other laws of physics. Applying the rule and developing the consequences allow us much detailed information about the emergent higher hierarchical levels. A whole array of new phenomena come into play that did not previously exist. We can move from the properties of atoms to the properties of molecules and collections of molecules.

This emergence is so intriguing because it leads us to the enticing question of whether, at higher hierarchical levels, there are not other nondynamical principles that introduce new kinds of behavior. Since all of chemistry emerges from one nondynamical rule, might there be another rule that will illuminate biology, or a rule that will give insight into cognition? The existence of the Pauli principle, which totally organizes the chemical world, is a powerful incentive to look for such rules at higher hierarchical levels. This may be a most useful heuristic. Many years ago, physicist Walter Elsasser suggested that there must be such a principle for biology. The search for biotic laws may be a search for pruning algorithms.

The Pauli principle indicates that our reductionist systems are not formally closed systems, and within science itself there is room for new kinds of pruning that will illuminate the emergent transition between hierarchies. This approach opens the way to thinking about problems. At each stage, any new selection principles can be subject to experimental verification, so that it is not a case of "anything goes." Nevertheless, new approaches are out there to be tested. There is an incompleteness in our current science.

Of all the emergence criteria, I find the Pauli principle the most encouraging in terms of eventually understanding higher levels. At any level there may be a presently unknown selection that will illuminate the hierarchical emergence in some way that we don't understand. That emboldens us to plunge ahead in our search for new laws of emergence that we have not dreamed of. If I were a betting man, I would suggest that emergence of mind will have at its deepest roots some such sort of selection principle.

This is all a bit frustrating, because I have no suggestions of how or where to look for a new rule, and must leave the task to the readers and others.

While the rules are quantum mechanics, and the selection criteria are governed by the Pauli principle and the formal emergence is the periodic table of the elements, all of chemistry emerges. At the root of this rule set is the covalent bond, which has been most extensively studied in the hydrogen molecule. Given two protons and two electrons and the rules noted above, the most stable states of lowest energy have the highest probability of the two electrons between the nuclei. This is the basis of the covalent bond, which exists between all nuclei where it is permitted by the rules of quantum mechanics.

This represents an enormous increase in possible complexity, for the 90 naturally occurring elements can now form into millions of possible chemical components that may exist in combinations, phases, and all the states of heterogeneous equilibrium. As long as the temperature is below 3000°, the evolving world is a world of chemical complexification.

Matter as we know it has emerged from the colossal explosion, the great condensing caldrons, and the enormous heat. These are repeated creations of the immanent God that follow from the laws of physics. Because of the Pauli principle, matter is informatic, and something akin to mind has already entered the universe. We now follow this through subsequent emergences. I repeat: matter is informatic.

## Chapter 7—Readings

Margenau, H. 1977 Reprint, *The Nature of Physical Reality,* Ox Bow Press (Contains a full discussion of the Pauli exclusion principle).

Pauling, L., and E. B. Wilson, 1935, *Introduction to Quantum Mechanics,* McGraw Hill.

Pauling, L., 1940, *The Nature of the Chemical Bond,* Cornell University Press.

# 8

# Planetary Accretion: The Solar System

The formation of second- and third-generation stars is appreciably more complex than the formation of first-generation stars. For, in addition to the primordial hydrogen and helium, the higher-generation protostellar cloud has many other atomic varieties, both as individual atoms, as well as in small particles that have condensed either gravitationally or by chemical attraction. These bits of matter have been designated as planetesimals or stardust. Because they have come from various parts of the cosmos and have been exploded off in various directions, they may have components of velocity at right angles to the line between them and the center of gravity of the protostar. As a result of the conservation of angular momentum, some of these particles will orbit the protostar, rather than being drawn into it. Therefore, in addition to forming a spherical star, the outer material forms a stellar system or, in the case we are most familiar with, a solar system. If one adds all the angular momentum vectors into a total for all the particulate debris, the material will orbit the protostar in a disc whose plane is at right angles to the total vector of angular momentum. The material may form into planets, asteroids, discs, or planetesimals, but it orbits in a disc-shaped envelope. An analog is the shape taken by a whirling mass of pizza dough. Newton's laws suggest that the orbits will be elliptical. Similar explanations will obtain for the rings of Saturn and moons of Jupiter.

The current view is summed up in a news article in the journal *Science* (1 October 1999):

> Star by star, these observations are providing physical evidence
> to support an old theoretical idea: that planets coalesce out of the

dust disks that surround many young stars. Researchers have discovered one star with both a disk and a planet, and other dust-enshrouded stars show features, such as gaps in their dust disk, that are "very suggestive, although not yet conclusive evidence for the existence of planets," says Ray Jayawardhana of the Harvard-Smithsonian Center for Astrophysics in Cambridge, Massachusetts.

Besides firming up the link between dust disks and planets, the findings are pointing to the crucial events along the way, together with a rough timetable for planet formation. A few million years after a star's birth, the tiny particles of dust encircling it rapidly coalesce into larger bodies and eventually into a handful of full-blown planets. After a few hundred million more years, the remaining debris crashes into the planets or is flung out of the system, ultimately leaving a relatively clean and dust-free planetary system like our own.

The process appears to be routine, cosmically speaking. After viewing hundreds of young stars, astronomers have found that many if not most are surrounded by these dust disks. So researchers now tend to believe that planets are the normal consequence of the birth of most stars—which would mean that there are billions and billions of solar systems hidden in the heavens. "It's becoming increasingly clear that the formation of our solar system is just one case of a general process accompanying the formation of a star," says Harm Habing of Leiden Observatory in the Netherlands. The only star we know close up is the sun, and it has an elaborate planetary system.

Present theories suggest that with the condensation of this protostar there were whirling disks that accreted and then mostly collected into planets. The process of accretion starts with atoms, molecules, and particles of star dust that have been ejected into space by novas, red giants, and other stellar explosions. The particulates are held together by chemical bonds. In studying this problem in 1952, Harold Urey emphasized that the formation of the solar system, outside of the sun itself, is in a temperature-pressure domain where the chemical properties of matter were as important as mechanical, gravitational, and magnetic factors. The emergence of the periodic table has produced interactions among atoms at a scale of $10^{-8}$ cm that influence subsequent emergences at a scale of the solar system. This interaction between the very small and the very large is a characteristic of emergent systems. Thus planet formation, geophysics, geochemistry, and celestial mechanics and, perhaps, biochemistry are a curious combination of the large scale and the atomic scale relating to the interaction of matter. The bits of space debris were presumably drawn in toward the protostar because of gravity, and the disk rotated because of the conservation of angular momentum. We are reporting a scenario for solar system formation that is generally accepted, but the details of the story are still being developed. It is part of the agendas of both astrophysics and geophysics.

The small pieces are drawn into planetesimals, asteroids, and other particulates. The governing large-scale rules are the laws of mechanics and the operative force is gravity, subject to conservation rules with respect to energy, momentum, and angular momentum. As the planetesimals get bigger, the collisions get more violent, due to the larger gravitational forces. Eventually a planet develops from the aggregation of the smaller pieces. The process doesn't happen all at once, and the collisions are often very violent as the gravitational force slams together the component pieces. The larger the protoplanet gets, the faster the speed of impact with smaller pieces. The resultant kinetic energy from each collision is converted into heat, raising the temperature, and often melting the surface.

The processes of impact and heating take place from the very first accretion to the impact craters after the planet is formed and continues to the present day as asteroids, comets, and meteorites impact the main planets.

Two kinds of processes take place due to these collisions. In the first, mechanical changes occur, altering the orbit and the angular rotations. In addition, the temperature rises due to a conversion of gravitational potential energy to atomic and molecular kinetic energy. There are three other sources of heat. First, as the planet condenses and compacts, gravitational potential is converted to heat. Second, radioactive isotopes carried in the planetesimals undergo decay, converting some mass energy of the isotopes to thermal energy. This is a continuing heat source for the maturing planet. Third, tidal effects from neighboring planets and the sun may convert celestial mechanical energy to heat. This collection of exothermic processes, in the case of the Earth, probably led to planetary meltdown and fractionation of the elements due to density and combining power among the elements. The latter depends, as we continue to stress, on rules generated from the Pauli principle. Also, the Gibbs phase rule, a principle of thermodynamics, governs formation of crystals and other states of aggregation or phases.

For the inner planets, the principal continuing sources of heat are solar radiation and radioactive decay. The intensity of the solar energy flux falls off as one over the square of the distance from the planet to the center of the sun. Recall that originally the planet is made of a combination of primordial hydrogen and helium and heavier elements cooked up in the stars and novae. Gas molecules that obtain a thermal velocity greater than the escape velocity will leave the condensing protoplanet, never to return. The average thermal velocity of a molecule is v = (square root of kT over m). T is the absolute temperature, k is Boltzmann's constant, and m is the mass of the molecule. This equation from kinetic theory relates the average molecular velocity of gases to the temperature. It is the kinetic theory definition of temperature. The condition for a molecule to escape is small mass and high temperature. Since hydrogen and helium have the smallest masses, they are the first elements blown away from the inner planets. Some hydrogen can be retained by being chemically combined with heavier atoms $H_2O$, $CH_4$, et cetera, but almost all of the inert helium escapes. Thus the inner planets such as Earth are largely made from heavier elements, and the larger, cooler outer planets such as Jupiter, with much higher escape velocities, are mainly hydrogen, the most abundant element in the initial condensation.

The problem of how the material in a solar ring condenses into a planet has not been completely solved. That reference to incompleteness in our understanding is, I fear, becoming repetitious in my statements, but I think it is a point worth noting. The good news for scientists is that there is much to be done; we will not run out of work. Returning to our forming planets, there are no violations of the conservation rules, nonetheless all the forces bringing together the pieces must be accounted for. Like many other parts of the puzzle, this aspect of planetary formation remains part of our work agenda. It also reflects our commitment to reducing astrophysics and geophysics to physics. The complexity may generate some emergent surprises.

While the formation of the moon is less certain, it now appears that a very large asteroid or planetoid impacted the Earth, knocking off a fragment that now orbits the planet. The Earth-moon system has a very large angular momentum so that, much like a spinning top, the direction of its north-south axis relative to the orbital plane is stabilized, and the weather tends to be regularized by the relatively fixed angle of the axis to the orbital plane. This is an unanticipated benefit of a lunar system in future emergences on Earth, such as life.

From the condensation of a second- or higher-generation star, a system then emerges that consists not only of the central star, but also a complex array of orbiting celestial material: planets, comets, asteroids, and smaller pieces. A solar system emerges. It is quite likely that the universe has a large number of such orbiting systems. The search for extrasolar planets is an active and exciting pursuit in present-day astrophysics. These planets or their satellites are the likely abode of any other life in the galaxy. Stars are too hot, and interstellar space is too cold.

Let's review the solar system in terms of general emergence concepts. The solar system is made up of about $10^{57}$ atoms, which are the agents at the chemical reductionist level. At the smallest scale they interact by interatomic and intermolecular forces. At the next level are larger-scale interactions of atoms defining phases and types of matter such as gases, liquids, solids, and dense plasmas. Then come large-scale gravitational and mechanical interactions.

The number of possible trajectories for the unfolding of the system is extremely large, difficult even to visualize. Certain selection factors lead to a planetary system. How general the selection factors are is really not known. If we could study many solar systems, as we may some day, we might then understand this emergence better and have a general theory of planetary formation. In part, our understanding of stellar emergence has come from the very large number of stars that we have been able to study in various stages of the stellar life cycle. With planets there is much less data.

One feature of planetary systems seems of particular advantage in the development of cosmic complexity. An aspect of nonequilibrium thermal physics is that the flow of energy from an energy source to an energy sink through a nonequilibrium intermediate system leads to the organization of the intermediate system. This is quite general. It is sometimes referred to as the fourth law of thermodynamics. The surfaces of planets orbiting stars are intermediate flow zones for two types of sources: the photon flux from the thermonuclear reactions in the star, and the energy generation from radioactive decay and gravitational compaction within the planet. In both cases, the energy sink is the near-absolute-zero (3° K) cold of outer space. Planetary surfaces seem optimal zones for chemical organization.

The emergence of planets in any case is part of the process of star formation, surrounding the stars with potential abodes for life. The rules with which they emerge are pointing to higher degrees of organization.

## Chapter 8—Readings

Canup, R. M., and Righter, K., Eds., 2000, *Origin of the Earth and Moon*, University of Arizona Press.

Morowitz, Harold, 1968, *Energy Flow in Biology*, Academic Press (Ox Bow Press Reprint, 1979).

Zerlik, Michael, 1993, *Conceptual Astronomy*, John Wiley & Sons.

# 9

# Planetary Structure

The material in a given orbit coalesces into a planet for reasons that are not completely clear and, in the case of the Earth, the planet probably underwent a meltdown. The energy for this process came from gravitational potential energy, kinetic energy of the components, heat of radioactive decay and radiant energy from the sun. Because of the small size of the Earth, the core temperatures were not high enough to be in the nuclear reaction domain, so that the fractionation of the planet was governed by chemical and mechanical factors.

We have discussed two types of nuclear reactions, and some elaboration might be helpful. The first is the nucleosynthesis that takes place in the core of stars and exploding novas and leads to the buildup of all the chemical elements that make up our universe. Among these elements, some are radioactive and decay into other components over varying periods of time ranging up to billions of years. This is the second type of nuclear reaction, often referred to as fission. Some of these radioactive isotopes condense into stardust and become part of the planet as it forms. The Earth thus captured uranium and thorium, which undergo a series of decays to stable isotopes of lead. The Earth also has potassium-40, which decays to argon-40 with a lifetime of 1.4 billion years. The decay of these isotopes gives off energy that usually ends up as heat. At present this is part of the convective heat flow that drives plate tectonics, and continental drift.

As a graduate student, I placed a thin-window Geiger counter into a bottle of ordinary reagent potassium chloride. It was startling as the clicks sped up into a buzz and the needle went off-scale. It is strange to think of all those clicks as a major component of the convective heat flow that drives the motion of continents, but so the calculations indicate. It's also scary to think of all the radioactivity within me from the potassium within my body.

Another feature of these long-half-life radioactive elements is that they can be used in radioactive dating, including the age of the Earth. Measuring the Earth's age has been a powerful factor in the argument between religious fundamentalists and those with a less literal interpretation of scriptures.

In the geophysical structuring of the planet following melting, a radial distribution of material took place, based largely on density of the elements. The inner core of the present Earth consists of a sphere of mostly solid iron 1,221 kilometers in radius. This is surrounded by a layer of iron nickel, a shell some 2,259 kilometers in thickness, known as the outer core. The core is surrounded by a 150-kilometer-thick shell designated the core-mantle boundary. The outer core is molten ferromagnetic material, and the Earth's magnetic field is believed to be due to electric currents in this part of the planet. The core is surrounded by a 2,500-kilometer-thick shell of silicates designated the mantle. The mantle itself consists of an upper and lower mantle with a transition region between them. In somewhat onion-like fashion, four more shells complete the planet: a 140-kilometer-thick asthenosphere, a 56-kilometer-thick lithosphere, a 21-kilometer-thick crust, and a 3-kilometer-thick ocean layer. (In some references, the crust and lithosphere together are called the lithosphere.) This shell conformation (or onion-like structure) is an emergent feature of the cooling planet.

The various zones engage in a number of dynamic activities. The upper mantle and outer layers sustain a convective heat flow from the rising thermal energy of radioactive decay. The lithosphere is divided into a number of plates that move, spread apart at rift lines and subduct crust at the boundaries. The process is known as plate tectonics and results in continental drift and recycling of materials, which precipitate on the ocean bottom, subduct into deep trenches, and recycle as volcanic magma and various geological uplifts.

A highly spherically structured and extremely dynamic planet emerges. There is little to suggest that any principles beyond thermodynamics, physical chemistry, and mechanics are required to understand the structure formation of the Earth shells, but nevertheless some details of Earth formation are still lacking. Even a very sophisticated macroscopic science such as geophysics does not present a complete explanation of all of the phenomena under study. There are still major disputes among the experts in almost all the historical sciences. We have a tendency in scholarly domains to overestimate the completeness of the theories. Some have called this the arrogance of the present. Even where the reductionist principles are thoroughly understood, the complexity of the unfolding leaves us with much to understand. This provides another role of emergence as a tool in understanding our world.

At this point, it is well to step back and note that we have viewed six emergences, each one has a reasonably widely accepted scenario of what happened, and there is much that is uncertain at each level. The pruning rules may be deep within scientific law, as in the Pauli exclusion principle, or may lie within the boundary conditions, as is likely in the shell structure of the Earth. We are only beginning to comprehend the different kinds of emergences. There has been a tendency on the part of those with an antitheological bias to assert without much reflection that each emergence is the result of a frozen accident. That is a philosophical-theological perspective, not a scientific statement in many cases. It is a dogma for some. I want to point out that the phrase "frozen accident" can be a bar to further study and possible understanding from the point of view of existing or newly developing theory. It is an assertion that there is no purposeful direction to the unfolding in time. But teleological direction may become an empirical question that therefore should not be answered *a priori*. I want to assert to my colleagues that an overly generous use of the concept "frozen accident" can have an anti-intellectual thrust.

There are general limits on the size of a planet. A planet that was too large would begin to heat up during accretion to the point where it would take on stellar properties such as nuclear reactions at the core. A very small planet would have such a low escape velocity that all volatiles up to quite high molecular weights would be quickly lost. The planet, made of high-atomic-weight elements, remains. The surface temperature of a planet is largely determined by its distance from its sun. Chemical bonds of any kind can only exist below certain temperatures. A very hot planet could not have stable molecular structures. A very cold planet would have reaction rates so low that over geological times no interesting chemistry could take place. There is at most a 500° K zone between too cold and too hot. In any case, mass, chemical composition including radioisotopes, and distance from the central star seem like the crucial factors in the surface organization of planets.

The chemical composition of the stardust depends on the laws governing fusion reactions and the history of the portion of the universe that contributed material. It is not clear what determines the amount of long-lived radioactive isotopes in the mix of planetary material, but it seems that these isotopes are necessary for tectonic activity.

The bottom line is that the emergent planet is complex, both structurally and kinetically. This is especially true if it has the character of an inner-solar planet. Within this complexity lie the possibilities of far more development of structures and processes and emergence of new features.

The experimental information on the shell structure of the Earth comes largely from seismography. There is a growing body of experimental data on the properties of materials under high pressures and temperatures. A new feature is the possibility of computer modeling of the kinetics of mass transport and chemical fractionation. Understanding the structure of the Earth seems to lie within the domain of normal physics and chemistry. It is, however, clearly a problem of great complexity and great difficulty. A lot of difference of opinion still exists among geophysicists, but one feels that this is a classical type of problem that will probably yield to the tools at hand. Of course, people who make such statements almost always are surprised.

## Chapter 9—Readings

Lunine, Jonathan, 1999, *Earth, Evolution of a Habitable World*, Cambridge University Press.
Rothery, David A., 1997, *Teach Yourself Geology*, Hodder Headline, London.

# 10

# The Geospheres

Since the abode of life on Earth is a thin spherical shell ranging from about five kilometers below sea level to five kilometers above sea level, we focus on the structure of this zone. Geochemists classify this extremely active outer region into four geospheres: the lithosphere, the hydrosphere, the atmosphere, and the biosphere. When the maturing planet settled down, the crust and the rocks on top of the crust hardened into the archean lithosphere. With time, three kinds of surface rock have emerged: igneous rock formed directly from the molten mantle, sediment formed by deposition, and metamorphic rock that comes about by the crystallization of other types. The lithosphere maintains cycles due to erosion and weathering, subduction and uplift, vulcanism and other processes. The surface is geologically dynamic. Even rocks do not last forever.

The hydrosphere consists of oceans, lakes, rivers, clouds, and other aqueous systems. It is formed largely from the water that outgassed in the melting of the Earth. The stardust that condensed in the Earth's formation contained various crystalline hydrates. Because of the vast amounts of hydrogen in the universe, any oxygen found had a high probability of ending up as water. Molecules of water can be tightly bound to minerals as hydrates. At high temperatures, the water is driven off as vapor and ultimately condenses into the hydrosphere.

In addition to the outgassing of water, various volatiles were driven off from the particulate matter during the meltdown of the planet. This gave rise to the atmosphere, a gaseous shell surrounding the Earth. The low-molecular-weight molecules of hydrogen and helium reached escape velocity and left the planet. The nitrogen, inert gases, and carbon dioxide became the atmosphere or the gaseous phase geosphere.

The biosphere emerged later. Although it is the smallest of the geospheres by mass, it is the most chemically active and hence plays a chemical role out of proportion to its size.

The importance of the biosphere was stressed by V. I. Vernadsky (1863-1945), who was one of the first to note that the character of the Earth's crust was due to the action of the biosphere. "Vernadsky's views on the importance of the biosphere are becoming increasingly relevant as we attempt to predict the climatic consequences of the greenhouse effect caused by the discharge of carbon dioxide and other gases into the atmosphere."

In our discussion of the Earth and its parts, we have tended to describe structures in an overly static way. We are dealing with far-from-equilibrium systems where almost all of the structures are dynamic. They are consequences of processes and would alter radically if deprived of the energy flows that maintain the forms. For example, consider a water fountain. It is produced by water under high pressure flowing through a jet and rising into the air then falling to the pool underneath. The fountain has a size and a shape, a height and a width. It is therefore a structure, but it is maintained by a process, the flow of water from the high pressure in the input pipe to the low pressure in the pond.

This fountain model of a dynamic structure characterizes almost all the objects we encounter in geochemistry and geophysics and indeed in biology, for they are all far from equilibrium. The hydrosphere contains all the bodies of water that drain into the sea or evaporate into the atmosphere. Atmospheric water precipitates. Dissolved and particulate material is carried to the oceans. Some of this material precipitates on the ocean floor and is subducted into the lithosphere. The oxygen of the atmosphere is the product of photosynthetic activity of the biosphere. The nitrogen of the atmosphere is a compartment of the biospheric nitrogen cycle, and the $CO_2$ of the atmosphere is part of the ecological carbon cycle.

All of the geospheres are dynamic and constantly exchanging material and energy with each other. The apparent division into geospheres is an emergent property of the complex array of processes by which the atoms and molecules of the planet interact with each other, under the influence of the energy flows.

The relative simplicity of dividing the geospheres into solid, liquid, gas, and living material serves to hide the complexity that lies behind all the processes and entities. Nevertheless, the limiting number of states of matter does serve as a pruning principle in the development of the Earth's surface.

After we have come to a better understanding of the emergence of the biosphere, we will be able to focus on the relations among the spheres, which will include the Gaia hypothesis, that relates the composition of the atmosphere to controlling activities of the biosphere.

Again let's note that the existence of three geospheres is a function of the size, temperature, and chemical composition of a planet. Small hot planets will not have an atmosphere because an atmosphere cannot be held by the weak gravitation competing with thermal velocity, which leads to escape. Large cold planets lack an atmosphere because any surface gas is pulled into a narrow layer in which the high gravitation force will compress it into a liquid or solid.

The existence of a hydrosphere depends on the presence of hydrogen and oxygen. Since these are among the most abundant elements in the present universe, we would expect water to be common. A three-phase hydrosphere consisting of ice, liquid water, and vapor requires a narrower range of conditions. Too cold results in a surface of ice; too hot yields an atmosphere of steam. The range of conditions for the coexistence of the three phases can be worked out in great thermodynamic detail.

A lithosphere requires the existence of silicates over a fairly broad temperature and pressure domain: it is not a universal planetary condition, and indeed is characteristic of the inner planets.

As for the biosphere (the fourth geosphere), the range of conditions and chemical composition is presumably much narrower. We will examine these in subsequent chapters. In any case, each planet will have a surface dependent on geophysical and geochemical constraints, and surface characteristics will vary for each planet. In a certain number of cases, conditions will be right for life to emerge.

A biotic planet will be radically different from a lifeless one. Most of the structure and complexity that we see around us are the consequence of life's having arisen and having become such a dominant factor: thus the large difference between Earth and our neighboring planets. It seems likely that continuing life requires a tectonically active planet so that life itself is a planetary property, or perhaps a solar-system property.

## Chapter 10—Readings

Faure, G., 1998, *Principles and Applications of Geochemistry,* Prentice-Hall.

Rankama, K., and Suhama, T., 1950, *Biogeochemistry,* University of Chicago Press.

Vernadsky, V. I., 1926, *The Biosphere* (English Translation 1998), Springer-Verlag.

# 11

# The Emergence of Metabolism

Toward the end of the accretion period, the Earth was still occasionally subject to large meteoritic impacts that boiled away the oceans into cloud layers that subsequently precipitated. The planetary surface, as we have noted, organized into a rocky lithosphere and a liquid hydrosphere. These were surrounded by a gaseous shell, the atmosphere. Beneath the lithosphere was a shell of hot magma that occasionally broke through the shell as a volcano or oozed out between spreading plates. Most of the surface was probably covered with water that constantly evaporated, collected in clouds and fell as snow, sleet, or rain. The atmosphere was largely nitrogen with small amounts of reductants such as hydrogen and variable quantities of $CO_2$. At this stage there was little or no oxygen in the atmosphere.

The three nonliving geospheres received chemical energy from two principal sources, the flux of solar photons that drove atmospheric reactions, and the energy sources within the core and mantle that led to the thermal division of chemical compounds into oxidants and reductants. We assume that the chemical activities resulting from these various energy flows resulted in the formation of the network of reactions for producing prebiotic molecules and ultimately protocells. Persistent life must be integrated into the chemical activities of the planet and requires a constant energy flow to keep from decaying to the thermodynamic ground state.

At this point a personal comment is in order, perhaps a mea culpa. The emergence of metabolism is the field of science that I have worked in for many years. Therefore, this section is somewhat more detailed than others that I am less familiar with. It is somewhat more judgmental, since I have had years to hone my opinion on the whetstone of the theories of others. My own approaches go back many years when I became enamored of the *Chart of Intermediary Metabolism*. I wrote the following:

The forthcoming interview by a reporter from a national news magazine was an exciting prospect for me as a young scientist. The interchange, however, went badly. Every time I mentioned adenosine triphosphate the reporter balked, and we had to go back to square one. The resulting article was garbled and gave a confused account of the research in progress. The lesson I learned was "stay away from big words and be careful what you say to the media." The troubling issue persisting over the years is that one of the great intellectual triumphs of all time is written in tongue-twisting polysyllabic words such as nicotinamide-adenine-dinucleotide-phosphate. How is one to tell the story of this important achievement in biochemistry to a general public unfamiliar with such language? One of the most significant advances in understanding the nature of life remains unknown to most people because it is inseparable from very long words that intimidate the uninitiated and keep them from insights that hold a wide range of unexplored implications.

The intellectual accomplishment that we here praise is not the work of a single individual, nor was it put together in a blinding flash of insight. Rather, it is the product of many researchers working at their laboratory benches over a period of more than a hundred years. Because the great structure came about so slowly and in such small steps, few biochemists have shown interest in extolling its magnificence. Their reticence may also come from the fact that the *Chart of Intermediary Metabolism* is very complicated and remains an unfinished edifice, like the great cathedral of Cologne, which was left with a crane still standing on one of its towers for many years as a symbol of the tasks for future generations. Instead of being the object of poetic rapture, the network of cellular reactions is groaningly memorized by biochemistry students. The enterprise for scientists usually centers on trees, or even branches, with complete disregard of a glorious forest.

But enter almost any biochemistry or molecular biology laboratory, and you are sure to find posted on the walls or doors printed sheets bearing a connected graph of all the major biochemical reactions that living organisms carry out in their cells. The chart is a great synthesis, a set of empirical generalizations summing up numerous experiments by generations of workers. It can be compared to other great achievements of the human intellect, such as the periodic table of the elements or the Linnean system of classifying species. While life scientists may be silent about the deeper significance of the metabolic chart, it stands in what amounts to a place of honor in almost every research center.

To envision intermediary metabolism in proper perspective, we start with the time-honored view of biology as unity within diversity. Variety and heterogeneity are clearly evident in the array of plants and animals that greet the eye whenever we take the time to look. Estimates of the number of extant species range from two million to ten million, and these are found in every habitat from deep oceanic trenches to the tops of high mountains. Diversity is one of those indisputable facts of life. Unity begins to emerge when we penetrate beneath the surface to the intracellular machinery and processes used by species to grow and reproduce. This examination at the microscopic level reveals common features of cell and organelle structure.

As we continue down the size range from organisms to molecules, we come to intermediary biochemistry, a collection of hundreds or thousands of enzymatic reactions by which a cell shapes matter and energy into forms appropriate for its own purposes. Here we sense the full impact of unity; the single *Chart of Intermediary Metabolism* applies with equal validity to all the millions of species that inhabit the planet. The core set of biochemical reactions of any organism from a bacterium to a great blue whale is found on the four-page chart lying on the table before me. No organism employs the full chart, but each species uses some substantial part of the reactions depicted. What may have been alarmingly complex to undergraduates studying for an examination becomes remarkably unifying and simple when we realize that it encompasses all of the diverse flora and fauna coexisting in the biosphere.

When a large body of knowledge is reducible to a compact system, scientists are tempted to look for a deeper law underlying the ordering. In the periodic table of elements, for example, the reasons for its form were ultimately found in the laws of quantum mechanics, particularly in the principle that restricts the number of electrons that may occupy each orbit in an atom. Once these laws were understood, it was possible to predict the periodic table in a detailed way from fundamental physical principles. At present there is no scheme for generating the metabolic chart from such basics, but hope springs eternal. And maybe, just maybe, there is a missing law that will resolve the basis of biochemistry, just as the quantum mechanical principles predicted an explanation of the major features of chemistry.

In the absence of a basic principle, we search for unifying features, and they are not hard to find. Almost every sequence of metabolic processes involves one or more reactions with molecules of adenosine triphosphate (reporter from my youth, please note). This ubiquitous substance, best known by the abbreviation ATP, is central to energy-processing in all cells. We find ATP printed along practically every line in the metabolic chart. If it were represented only once in our network of biochemical pathways, the drawing would be a giant rosette with all lines passing through the center. ATP is the final energy-transfer molecule in almost all cellular processes. The reactions of this compound heat our bodies, power our muscles, charge our nerves, and otherwise drive the processes of life.

Accompanying ATP is a series of other substances that play major roles in energy transfer. Each contains the molecule adenine built into its structure. In the language of life this atomic configuration appears as the symbol for an energy storage molecule, yet the adenine portion itself plays no part in the energy process. The whole idea seems information-rich, somehow rather too linguistic or poetic for the grind-and-extract business of biochemistry; yet there it is. In addition to being a signal for energy transfer, adenine also constitutes a major symbolic component of the genetic code, being one of the four bases of DNA and RNA. Can there be some deep and fundamental, yet hidden, relationship between coding and energy transfer? It is a question worth addressing, for an understanding of adenine seems to lie close to the biochemical secrets of life.

Beyond the particular characteristic of the ubiquitous adenine, many less-sweeping propositions emerge from a study of the metabolic chart. These, too, stand as challenges to biophysicists, biochemists, and biophilosophers, urging them to penetrate deeper into the relations standing behind the experimental facts. Deriving the grand structure from a more primitive principle would give great insight into that perennial, yet ever significant, question: What is life? It would also spare another generation of students from having to memorize the entire chart.

With the above in mind, we think about how the chart emerged from the vast world of the chemically possible.

The emergence of the biosphere from the operations of the other three geospheres, which we recognize as the origin of life, must have involved imposing constraints in two spaces or domains. At some point, physical space in which metabolic systems function must be limited to keep reactants from diffusing away. This can be accomplished by adsorbing material on surfaces, trapping it in interstices or enclosing it in vesicles. The second kind of space, mathematical in nature, is an information space where the dimensions specify the properties of molecules and the connections specify chemical reactions in a graph-theory representation of chemistry. This abstract space follows from the rules of organic chemistry. Confinement in information space determined by what is chemically possible might come before or after limiting the physical space.

The two types of constraints in this space and phase space resulted in selection rules for emergence. Network autocatalysis consists of the end product of a reaction sequence entering into its own synthesis so that the rate at which a molecule is produced depends on how much of it is present. Network autocatalysis, which requires an entire array of small molecules, provides a self-organization of the kinds of molecules that can be generated. Template catalysis, which requires large structures, also alters the kinetics and the molecular distributions. Thus the numbers and types of molecules present is an emergent property of the reactions in chemical phase space. If polar molecules with hydrophobic and hydrophilic ends are synthesized and they form membrane vesicles, then the molecules that are generated in the system are captured within the protocells subject to the permeability of the membrane.

The prebiotic chemical organization of the planet was most likely carried out at two kinds of locations. The first was the surface layer where sunlight, including short-wavelength radiation, drove photochemical reactions and produced a variety of product molecules. The second site of activity was beneath the oceans where at certain rifts under high temperature and pressure the ascending magma hit the water and a variety of redox reactions synthesized a series of product molecules. This notion of two sites follows from the necessity of having an energy source to drive the process.

We do not know which energy source was the primary driver for life's origin, but clues exist for examining certain generalizations from the universal intermediary metabolism of contemporary living forms, the chart we have discussed above. This focuses on prokaryotes such as bacteria, which are believed to be the earliest organisms. So in the emergence of cellular life on the planet, we assume a sequence: metabolic-like processes, protocells, prokaryotes, and the most complex cells, eukaryotes. Among prokaryotes a distinction is made between autotrophs, which require no environmental organic compounds and get all their carbon from simple one-carbon molecules such as carbon dioxide or methane, and heterotrophs, which have such requirements. We assume the first cells were autotrophs, since heterotrophy imposes enormous requirements on the kinds of carbon compounds in the environment, an unreasonable demand on the nonbiological geospheres.

There are two classes of autotrophs, photoautotrophs and lithoauto-trophs. The former get their energy from light, and the latter from oxidants and reductants in the environment. Oxidants are molecules such as oxygen or sulfur that acquire an electron in a reaction, while reductants are mol-ecules like hydrogen that give up an electron in a reaction. Chemoauto-trophs are the simpler of the two, because light harvesting and conversion of photon energy to chemical energy is a more complex process and re-quires a number of highly specialized molecules. Lithoautotrophs get their energy from oxidation-reduction processes, and we shall assume these were the earliest types of cells.

Because all species of prokaryotic autotrophs have a large core segment of the reactions of the metabolic chart in common, we assume that this must go back to the earliest biosphere or before. The chart of intermediary metabolism is thus a virtual fossil of the earliest biochemistry. At the very core of this metabolism is a reaction network known as the Krebs cycle, after its discoverer, Hans Krebs. It is also known as the citric acid cycle or the tricarboxylic acid cycle (four of the components—citric acid, cis-iconitate, oxalo-succinate and isocitrate—are tricarboxylic acids). We know it is the metabolic center, for out of this cycle operating in autotrophs come all the pathways to sugars, fats, and amino acids. And from these come the nucleic acids, vitamins, and cofactors. Thus all synthesis ulti-mately comes from this cycle. Organisms that lack these steps must eat organisms or the products of organisms that possess these steps. Thus the total chart of intermediary metabolism must characterize every ecosystem.

A new variant of the Krebs cycle, discovered in the 1980s, is called the reductive TCA cycle. As carried out by the thermophilic (heat-loving) au-totrophic bacteria of the genus *Hydrogenobacter,* the citric acid cycle is driven in the reverse direction of the conventional oxidative cycle by re-actions of oxidants and reductants in the environment and serves as an instrument of synthesis to incorporate $CO_2$ and to synthesize citric acid, acetic acid, pyruvic acid, oxaloacetic acid, malic acid, fumaric acid, suc-cinic acid, alpha keto glutaric acid, isocitric acid, and other components. This network in *Hydrogenobacter* is autocatalytic. The product molecules increase the rate of their own synthesis. If it could have taken place without enzymes, it would have served as a sink for environmental carbon, col-lecting molecules of $CO_2$ into organic molecules that are the starting points for further stages in the metabolic production of biochemical molecules. This could have been the beginning of the kind of information constraints we referred to above.

If autocatalytic networks are at the beginnings of biosynthesis, their components could have been the first emergent structures made of organic molecules, selected by a set of rules of organic chemistry governing the reactions accompanying the flow-through of energy from the high-frequency sources (photons or redox couples) to the low-frequency thermal sink, such as the cold of outer space. Thus the primordial reductive citric acid cycle might have been a very early emergence of metabolism on the route to biogenesis.

The original space enclosures can come from the environment or can themselves emerge from the prebiotic metabolism producing membrane-like molecules. In the case of the latter, the logic of self-replication will be fulfilled, and we will have a protocell capable of reproducing itself as an entity. Thus if we start with acetic acid in the autocatalytic cycle, we note that it functions in the cycle by being converted to pyruvic acid. Another reaction is possible in which acetic acid is converted to malonic acid. This is the first step on the way to fatty acids such as stearic acid.

Stearic acid is the kind of amphiphile that can form into a planar polar lipid bilayer such as is found in biological membranes. These planar sheets can form into vesicles that will grow as more stearic acid is synthesized. As the metabolites increase and the membrane grows, the protocells will spontaneously divide. These are self-replicating entities; they represent the emergence of the biosphere. This last and smallest of the geospheres will grow increasingly important as it develops increasingly specific chemical activity and becomes a prime catalyst for much of the geochemistry of Earth's surface.

The view of the emergence of biochemistry that we have been discussing represents a paradigm shift from what the reader may have encountered in biology courses where it was assumed that random products of free-radical reactions lead to monomers, then to polymers, then to cells. In the view elaborated here, selection rules lead to a core metabolism that then produces an ordered hierarchy of emergent structures and functions. These become increasingly complex, leading to the sophisticated chemistry of the universal ancestor. This is a very different view than may have been taught in standard introductory courses, but I believe that it is a much more probable scenario, owing to what we have learned in the last 30 years.

It appears that the chart of intermediary metabolism arises for a collection of atoms of the periodic table interacting by the rules of organic chemistry under the right conditions. The ultimate emergence of metabolism seems embedded in the laws of chemistry, but the reactions are a tiny subset of all possible organic reactions. We must search for the pruning algorithm.

## Chapter 11—Readings

DeDuve, Christian, 1995, *Vital Dust: The Origin and Evolution of Life on Earth*, Basic Books.

Fry, Iris, 2000, *The Emergence of Life on Earth*, Rutgers University Press.

Morowitz, Harold J., 1992, *Beginnings of Cellular Life*, Yale University Press.

# 12

# Cells

The emergence of self-replicating protocells marks a major transition in the evolving world. First, replication of similar objects populates the world with distinguishable entities. There is thus a system memory since the replicated objects resemble their progenitors. This property of systems could even have occurred before macromolecules, as the memory could be embodied in a collection of small molecules and the reaction network among them that keeps regenerating itself. Second, variation or mutation populates the world with variety. Third, competition among the variety of forms selects for fitness. Fourth, complexity and emergence generate continuing novelty. The relatively simple protocell gives rise to all the subsequent novelty that comes forth in the living world, which some would think of as emanating from the mind of the immanent God. It all follows from the laws of nature and selection for the reified from the domain of the possible.

In this realm, novelty piles upon novelty, and we change from the rule systems of physical chemistry to those allowable rules of biology. With the emergence of distinguishable competitive protocells, the world becomes Darwinian, and we move from the domain of relative simplicity to the kind of complexity that eventually leads to the emergence of mind. Later we will deal with the inevitability of complexity. This chapter deals with the evolution of chemical and structural sophistication, moving from protocells to prokaryotes.

As a guide to these emergences, we again focus on the metabolic chart and the guiding principle that metabolism recapitulates biogenesis. That is, we assume that the smaller the number of reaction steps from $CO_2$ to a given metabolite, the earlier that metabolite occurred in the development of life. Thus, for organisms using the reductive tricarboxylic acid cycle, the earliest steps were the synthesis of the core chemicals acetate, pyruvate, oxaloacetate, malate, fumarate, succinate, alpha keto glutarate, oxalosuccinate, isocitrate, and citrate. These are listed not to impress the reader with the jargon, but to note how few chemicals there are at the core of metabolism.

Because of its central role in membrane formation, the pathway from the core cycle to lipids, or fatty molecules, seems like an appropriate starting point. This is done presently by a series of repetitive reactions that require an elaborate enzyme system. But the reactions are quite basic. Acetic acid plus $CO_2$ goes to form the three-carbon malonic acid. The malonic acid adds two of its carbons to the existing chain, making it longer. A series of reactions eliminates oxygens to produce saturated fatty acids. This builds up two carbons at a time, leading to the fatty acids $CH_3(CH_2)_{2n}COOH$.

These molecules are examples of a class of structures called amphiphiles, meaning love of both kinds. The both kinds are oil and water. One end of the molecule is an alkane chain that is attracted to oils, since it is at a lower chemical energy when dissolved in oils. Such structures are called lipophilic. The other end is an organic acid that is at its lowest chemical energy when dissolved in water. Amphiphilic molecules in water cluster in a number of forms with the lipophilic ends of the molecules in contact with each other. The other ends called hydrophilic are immersed in the surrounding water. Particulate arrays of this type are called colloids. Many kinds of clusters are possible. This is the basis of many of the phenomena of the chemistry of arrays of molecules.

When the lipophilic moieties are in the appropriate size range, the colloidal aggregates are in the form of planar sheets with the oily core sandwiched between two hydrophilic planes. Under appropriate conditions, the planar structures can fold into spherical shells in the size range of living cells. The formation of membranous vesicles is generally spontaneous. It is the lowest energy state of aggregates of these molecules under these circumstances. This is exemplary of how physical chemistry leads to cell structure—in this case, the membrane.

The contemporary prokaryotic cells, which may not have changed much in four billion years, have the following components:

1. Membranes
2. Cell walls
3. Ribosomes

4. Enzymes
5. Closed loop genomes
6. Transfer and messenger RNAs
7. Cofactors

One question to be raised is how the prokaryotes emerged from the pro-tocells in a rather short period of time. The process of going from small-molecule chemistry to organelles works by synthesizing small molecules and then making structures out of these molecules. In general, this is done by three methods:

(a) The first has already been discussed. The molecules aggregate into a new phase that has novel colloidal properties based on such factors as dielectric constant and solubility. This is how membranes are made.
(b) The second method starts with molecules that all have in common A and B ends and can form chemical bonds of the following type: A—BA—BA—BA—B.
These long structures are linear polymers of the kind found in proteins, sugars, and nucleic acids. The chains of monomers may then be folded in various three-dimensional structures. The use of regular linear polymers is a form of modular construction that greatly simplifies the chemical task of building large structures from smaller units.
(c) The third uses linear polymers that are cross-linked in various ways. Many polysaccharides form in this way and may give rise to rigid structures such as cell walls.

A group of polymer-forming compounds that are immediately synthesized from the reductive citric acid cycle intermediates are the amino acids. Thus, keto acids in the presence of ammonia and reductants form amino acids. We then have:

pyruvate→ alanine
oxaloacetate→ aspartate
alpha-keto glularate→ glutamate
and, with further additions of ammonia,
aspartate→ asparagine
glutamate→ glutamine.

Thus, 5 of the 20 universal amino acids are made directly off of the citric acid cycle in one or two steps. The other amino acids are made by a series of reactions starting from these five. A linear polymer of amino acids is called a polypeptide, which is the basic structure of proteins.

A typical amino acid consists of four groups attached to a carbon atom. These are a hydrogen atom, an organic acid group (COOH), an amine group ($NH_2$), and a side chain. The side chains are very variable and give the individual amino acids their special characters. The acid end of one amino acid group can combine with the amino group of another amino acid, to form a peptide bond that consists of two amino acids held together by covalent connections C-N-C and give rise to a dipeptide. Since most enzymes are protein catalysts made from linear polypeptides, template catalysis is an emergent property of such arrays of amino acids. The existence of this template catalysis allows the system to carry out, with specificity, more sophisticated chemical reactions than were previously utilized.

Regarding amino acids as agents and using the rules of polymer formation or the formation of peptide bonds, the array of amino acid sequences in the linear polymers produces a vast number of possible polymers. For example, starting with the 20 naturally occurring amino acids, a chain of 100 amino acids can be one of $20^{100}$ possibilities. In terms of power of 10, this is greater than $10^{101}$(the number one, followed by 101 zeros), which is truly a huge number. Many of these sequences will be catalytic for a wide array of reactions. If these catalytic sequences are somehow selected for, then enzymatic catalysis is an emergent property of the whole system, which is somehow different from the individual components. In a similar manner, polymers of ribonucleotide monophosphates with catalytic activity can form, thus allowing another pathway to catalytic synthesis. From protocell to prokaryote is thus a series of emergences.

All amino acids except glycine can exist in pairs called stereoisomers, which are mirror images of each other, one called L (levo) or left-handed, and the other called D (dextro), or right-handed. One of the enigmas of biochemistry is that all amino acids that are coded and incorporated into proteins are of the L configuration. This too turns out to be an emergent property.

All amino acids in autotrophs are synthesized along pathways in which they get their amino group ($NH_2$) from glutamate. The reaction is transamination, which takes place by a process called the ping-pong reaction. There is an intermediate called pyridoxal phosphate, which is a derivative of vitamin B6. The reaction is:

**Ping** {glutamic acid + pyridoxal phosphate → oxaloglutaric acid + pyridoxal amine phosphate} (The amine group transfers to pyridoxal phosphate)

**Pong** {pyridoxal amine phosphate + keto acid → new amino acid + pyridoxal phosphate} (The amine group transfers to keto acid)

Due to certain chemical structural rules, the new amino acid will have the same steric structure as the glutamic acid. Thus a selection rule in the stereochemistry leads to a generalization that, if glutamic acid is L, all amino acids are L.

These features and many others participate in the pruning of all possible chemistries to a restricted metabolic chart. The three-dimensional structure of proteins depends on all amino acids' being of the L form, or in any case of the same form.

In the same sense that amino acids are in the L configuration, naturally occurring sugars are in the D configuration. In cells that operate by the reductive TCA cycle, sugars are produced by this pathway:

$$\text{acetate} + CO_2 \rightarrow \text{pyruvate} \rightarrow \text{phosphoenol pyruvate}$$
$$\rightarrow 2 \text{ phosphyglycerate} \rightarrow$$

This pathway leads to D-glyceraldebyde 3 phosphate, which is the source of the D chirality in all sugar synthesis.

Sugars can also form linear polymers using the same kind of dehydration reactions that lead to polymers of amino acids. Sugar polymers can form cross-links and give rise to structures of considerable mechanical strength.

Consider cells containing metabolites and polymers. As polymers pile up inside the protocells, the osmotic pressure will rise internally relative to the outside, and the cell will be in danger of blowing up. This was solved historically by laying down a cross-linked polysaccharide cell wall outside of the membrane. The wall restrains the membrane from expanding, breaking, and releasing the contents of the cell. Almost all modern-day eubacteria have a wall, except for those that have subsequently lost it evolutionarily (such as the mycoplasma). These cells have solved the problem of stability in another way. They tend to be parasites that take cholesterol from their hosts to stabilize their membranes.

The primary molecules at this stage of emergence are fatty acids, intermediate metabolites such as keto acids, amino acids, sugars, and polymers of the above. Starting from sugars, amino acids, and carboxylic acids, pathways emerge for the synthesis of nitrogen-containing heterocycles. These are planar molecules, which turn out to be appropriate for information-carrying templates. They can be attached to other molecules to give rise to linear polymers such as ribonucleic acid or can be components of molecules such as coenzyme A, which take part in fine-tuning chemical reactions in the cell. A set of processes then emerges whereby sequences of amino acids can be encoded in sequences of nucleotides. This leads to a stable hereditary scheme. Coding is clearly an emergence of greatest significance. It leads to genes and the stability of genetic information for many generations.

Among the various structures, large macromolecular arrays of proteins and ribonucleic acids (ribosomes) emerged and serve as scaffolds and enzymes for the synthesis of proteins specified by polymers of ribonucleic acid designated as messengers. Ribozymes, RNA enzymes, also emerged.

Cellular information is further stabilized by storing the nucleotide sequence in double-helical DNA sequences or genes that are strung together in a closed loop or chromosome. In all the cases we have discussed, a novelty of chemical structure in making a new small-molecule level permits macromolecular changes with major structural and functional consequences. Biology moves from small-molecule chemistry to structure.

The prokaryotic cell has emerged: the universal ancestor. We have rather rushed through the emergence of prokaryotes, which could have been divided into several emergences. This lack of detail is due to lack of knowledge. The intermediate forms did not survive; however, the understanding of the intermediate emergence lies within the domain of macromolecular chemistry, and what cannot be found in the fossil record is subject to experimental and theoretical study in the world of chemical networks, physical study of macromolecules, and a better understanding of system properties.

On the details of these issues scientists are deeply divided. If one believes that the steps from protocells to prokaryotes compose a series of frozen accidents, then we shall never recover the processes and can only study the end states. If one believes that these emergences are rule driven and highly deterministic, then we can look forward to an ever-increasing understanding of how the laws of the universe, including the laws of emergence, have led to the molecular biology that we have come to know and understand.

It seems likely that in nature the transition from protocells to prokaryotes took place in less than 200 million years. It may have been far, far faster than that.

The central theme of prokaryotes is the synthesis of macromolecules and the use of these molecules for structure and function. The problems of building structures with specific functions are architectural, chemical, and informational. By architectural, we mean the three-dimensional structures of the modular units and the mode of assembly. The structural features are, at their basis, chemical; bond distances and bond angles. An example is the peptide bond between two amino acids, which we have discussed. The bond is planar, the C, O, and N lie in the same plane. The reason for this is ultimately chemical, but it is involved in the three-dimensional structure of all proteins.

The active site of enzymes is determined by the three-dimensional structure and is the site where the chemical reactions take place. The reactions are determined by chemical groups at the site. It represents an overlap between architectural and chemical features of macromolecules.

Because chains of amino acids and nucleotides carry information and the two kinds of messages relate to each other, proteins, DNA, and RNA are all informational.

The macromolecules are the subunits out of which larger units, such as ribosomes and enzyme clusters, are built. Once again, architecture, chemistry, and information come together.

Why do we treat the transition of protocells to prokaryotes as a single emergence? That is a valid question. There is, as we have noted, a certain arbitrariness in counting emergences. The more we know, the more we can fine-grain our treatments. This supports our contention that the unfolding of life involves many, many emergences, and this multiplicity is part of the nature of the story we are trying to tell. There may also be a biotic principle of the nature of the Pauli principle that will make the transition more understandable.

We tend to think of emergences as a post-computational modeling view of biology. In the last chapter we noted a remarkably prescient book written in 1964 entitled *The Emergence of Biological Organization*. The author, Henry Quastler, developed a series of emergences from probionts to prokaryotes. Today the book appears surprisingly contemporary in outlook.

We are just at the beginnings of science. The unknown is not unknowable. I believe that we should only postulate emergence by accidents when all other explanations have failed, and that may require some patience.

At the levels of cells we move from Cassirer's world of physics to his domain of biology. Once formed, the biosphere joins the other three geospheres and has a profound effect on geochemistry. This effect of the biosphere on other geospheres is often embodied in the Gaia hypothesis, which deals with the role of the biosphere in regulating the other geospheres and being regulated by them. An example of this kind of control is the oxygen content of the atmosphere.

Almost all the oxygen in the atmosphere is the result of the photosynthetic dissociation of water into free $O_2$ and biological hydrogen in reduced compounds. Oxygen is produced by photosynthesizers and used up in aerobic metabolism. Let's examine why oxygen in the atmosphere is about 20 percent of the gases. If the concentration grew much larger, spontaneous combustion such as forest fires and other oxidations would drop the oxygen level by using it up. If, on the other hand, oxygen dropped much below 20 percent animals' metabolism and oxidative processes in heterotrophs would decrease. This would use less oxygen, and the amount in the atmosphere from photosynthesis would rise. Thus the biosphere regulates the oxygen content of the atmosphere. This is a simplistic view of what is called the Gaia hypothesis.

At the emergence of cells, we are also in the domain where the world becomes Darwinian, for different kinds of cells will compete with each other, and the molecular memory of what generated fitness will flourish.

## Chapter 12—Readings

Henderson, Lawrence J., 1913, *The Fitness of the Environment,* The MacMillan Co.

Lovelock, J. E., 1979, *Gaia,* Oxford University Press.

Morowitz, H. J., Deamer, D. and Heintz, B., 1988, *The Chemical Logic of a Minimal Protocell/Origins of Life Evol. Biosphere* 18, 281–287.

Quastler, Henry, 1964, *The Emergence of Biological Organization,* Yale University Press.

Smith, John Maynard and Szathmary, Eörs, 1999, *The Origins of Life,* Oxford University Press.

## 13

# Cells with Organelles

Along the evolutionary pathways, there are points where splits appear in the tree of life that shape all subsequent development. In this chapter we will first discuss such a bifurcation between the two kinds of prokaryotes that have emerged, archea and eubacteria. Subsequently, in another branching, the eukaryotes formed from the prokaryotes by various processes of combination of members of archea and eubacteria. The unicellular eukaryotes, the protoctista, then gave rise to the multicellular eukaryotes, the plants, fungi, and animals. Somewhere along the later evolution of animals, a split occurs, giving rise to the protostomes and deuterostomes. The first of this group of major evolutionary divides is deep within the prokaryotes or bacteria. These are the simplest extant cells, and their universal features probably preserve aspects of the earliest prokaryotes.

Fairly soon after the origin of replicating cells, a cell architecture and metabolism, that of the prokaryote, evolved and locked in place in the universal ancestor. This scheme of living function has been enormously successful, having lasted for four billion years. Even though other cell forms have emerged from the prokaryotes, the original plan is still there and still competing successfully. The basic metabolism has become universal. The scheme functions using reactions of the universal metabolic chart that we may regard as a four-billion-year-old virtual fossil. Let us review the cell components. The cells have an amphiphilic bilayer membrane surrounded by a cell wall for osmotic integrity. Amphiphiles are molecules with an oil-loving end and a water-loving end. The genome is a closed loop of double-stranded DNA, from which messenger RNA is transcribed. There are ribosomes made of both RNAs and proteins that are the sites of protein synthesis. There are protein enzymes in the cell membrane and the cell matrix. This form and function have been enormously successful. Again, we may ask, is it a series of accidents or a uniquely successful solution to the requirements of cell biology?

Two different forms of the prokaryotic cell plan have emerged: eubacteria and archaea. They are sufficiently similar so that it is clear that one has evolved from the other, or both have evolved from a common ancestor. The two differ at a chemical level in the sequence of ribosomal RNA and the structure of the ribosomes, the chemical nature of membrane lipids and method of attachment of lipids to the polar groups, structure of the cell wall, and nature of the DNA polymerase. The differences between these two major kinds of bacteria have separated these taxa for billions of years.

The theory (in its more general form) of how eukaryotes arose from prokaryotes has been developed in its modern form by Lynn Margulis. She states:

> All protoctists evolved from symbioses between at least two different kinds of bacteria—in some cases, between many more than two. As the symbionts integrated, a new level of individuality appeared.
>
> Many different combinations of ancient bacteria into symbiotic consortia did not pass the test of natural selection. But those that survived gave rise to modern-day protoctist lineages which may be grouped according to their organelle structure.

Protists evolved by one cell engulfing another and the two living in symbiotic relationship. As they adapted to this symbiosis, redundant functions were lost and the evolving organisms changed their character. Thus chloroplasts have come from cyanobacter, and mitochondria have come from aerobic bacteria. The entire theory is elegantly summed up in the Margulis book *Symbiotic Planet*. Eukaryotic cells, a totally new life form, emerged.

A consistent but somewhat more elaborate scenario for the origin of eukaryotes has been presented by John Maynard Smith and Eörs Szathmary. They start with the view of Tom Cavalier-Smith that the earliest organisms were the eubacteria. Occasionally such cells are subject to catastrophic loss of cell wall. This leaves the cell in danger of blowing up under the force of internal osmotic pressure. There are at least three solutions to this problem. The first is to develop a stronger and more flexible membrane by a change in the nonpolar molecule and mode of attachment. This led to the archea. The second is to develop a more stable membrane and an internal apparatus to provide a skeletal structure for the membrane to attach to. This led to the eukaryotes with microtubules as the interior framework. The third and much later method is a strengthened membrane and very small cells. This is the method of the mycoplasma.

In any case, the Cavalier-Smith view is that the eubacteria by loss of cell wall led to the archea and the proto-eukaryotes. From there on, the Smith and Szathmary scenario is similar to that of Margulis, but in some aspects more detailed.

At this point let us look briefly at the characteristics of prokaryotes and eukaryotes. While there is a great variation in cell size, eukaryotic cells are on the average about 1,000 times the size of prokaryotic cells, which lack membrane-bounded organelles. Eukaryotes almost universally have mitochondria and membrane-bounded nuclei as well as species-specific chloroplasts and other assorted structures, including Golgi bodies and centrioles. The eukaryotic cell is an order of magnitude more complex than the earlier prokaryotic cell.

The endo-symbiotic theories of major evolutionary change give a new perspective. In classic evolutionary theory, the tree branches by mutation or other internal modification. From the symbiosis perspective, new forms emerge from a coalescence of existing types; thus, eukaryotes emerge from a coming together of prokaryotes. Fitness is achieved by merging separate pathways of fitness. It is an extremely efficient mode of developing new evolutionary forms. In the genetic recombination associated with sex, this idea of emergence by joining functional solutions has been built into the reproductive scheme of the eukaryotes. But it is entirely genetic, as opposed to endosymbiosis. A current example of endosymbiosis in progress is the green hydra, animals that are green because of intracellular algae that provide food photosynthetically.

It is interesting to speculate why this endosymbiosis program took one to two billion years to emerge. Recall that one of the properties of prokaryotes is a rigid cell wall to keep the cell from osmotically exploding from the accumulation of molecules in the interior. The wall is a barrier to exchange of material between cells or to one cell engulfing another. Developing alternative solutions must have taken a long time.

In spite of the cell wall barrier, prokaryotes have discovered ways to exchange genetic information. In bacterial transformation, DNA released by one cell is taken up by another cell and incorporated in the genome. This usually occurs between closely related taxa, but the phenomenon allows for merging of information molecules across the taxa and through the cell wall, and hence for emergence by merging. Other methods of transferring DNA have arisen including bacterial mating and viruses conveying some bacterial DNA from cell to cell. However, as long as both cells were walled, transfer was restricted to macromolecules, as contrasted to organelles.

There are, as noted, in principle three ways in which a cell can prevent osmotic lysis: using a rigid restraining wall, developing a strong elastic membrane, or building an internal framework. If we look at current organisms, eubacteria, fungi, plants, and some protoctists have rigid cell walls, while animals, and some other protists lack cell walls. The latter have either incorporated into their membrane lipids, sterols such as cholesterol, which provide the necessary strength and elasticity to the membrane, or used a different type of amphiphile. Special lipids are required for the stability of cells without walls.

In one case, the synthetic pathway from acetate to cholesterol takes about 30 enzymatic steps, so it may indeed have taken a long evolutionary route to the chemicals that would have produced stable wall-less cells. A look at another route from walled to wall-less cells is instructive. The mycoplasma are, I believe, the only prokaryotes that are wall-less. Nucleic acid analysis indicates that they are derived from walled bacteria that adopted a parasitic existence and lost a lot of chemical pathways including wall synthesis. They achieved the necessary stability without a wall by taking sterol from their hosts and using it to make a sterol-containing membrane that permitted wall-less existence. Hence the cholesterol in the membranes of most species of mycoplasma. Stability is also achieved by very small cell size.

The first two billion years gave rise to many kinds of prokaryotes adapted to the vast array of niches available on the Earth. One of the most common groups of prokaryotes is cyanobacter, photoautotrophs that use sunlight to convert water and $CO_2$ to sugar and oxygen. At some point, cyanobacter were presumably ingested by proto-eukaryotes and adapted to living within the cell. They continued their unique role of photosynthesis and continued to divide in coordination with cell division. They progressively lost functions associated with independent existence and became replicating cell organelles providing the cell with chemical energy from sunlight and using the host cell for other nutrients. They became organelles, chloroplasts.

The origin of the membrane-bounded nucleus and the process of meiosis is less clear but is associated with the emergence of the eukaryotes and became a universal feature of the protoctista. The protoctista are a broadly diverse group of organisms described by Margulis and Schwartz in 1998 as:

> Nucleated microorganisms and their descendants, exclusive of fungi, animals, and plants, evolved by integration of former microbial symbionts. Non-meiotic or meiotic with variations in the meiosis-fertilization cycle. Fossil record extends from the Lower Middle Proterozoic era (about 1.2 billion years ago) to the present.

They are characterized by many inclusions or organelles per cell, and the rich variety of organelles led to a great spread of speciation.

Protoctista can obtain energy and nutrients in three general ways. Primary producers carry out photosynthesis and require $CO_2$, a source of nitrogen, and other soluble nutrients. Saprophytes absorb soluble molecules made by autotrophs or decomposing organisms of all types. Animal-like protists ingest other organisms and then digest them to obtain nutrients.

Photosynthesizers, because they do not need to ingest particulate matter, revert to one feature of the ancestral prokaryote form and develop a rigid protective cell wall. Saprophytes absorb nutrients by diffusion and also do not require ingestion of particles. They also evolve a protective cell wall. Animals continuing to ingest particulate food do not develop cell walls and continue through evolution with cell membranes reinforced with sterols and an internal microtubule structure.

Protoctists emerged from the prokaryotes. They have much of the chemical complexity of their progenitors and additionally have the complexity of symbiosis embodying all the combinatorial ways of putting together a variety of organelles. To ask if evolution involves increasing complexity, it is only necessary to compare the protoctists with the prokaryotes. In short, a new level of complexity has emerged. Smith and Szathmary sum up this emergence in the following way:

> Although we have written of the origin of the eukaryotes as one of the "major transitions," it was in fact a series of events: the loss of the rigid cell wall, and the acquisition of a new way of feeding on solid particles; the origin of an internal cytoskeleton, and of new methods of cell locomotion; the appearance of a new system of internal cell membranes, including the nuclear membrane; the spatial separation of transcription and translation; the evolution of rod-

shaped chromosomes with multiple origins of replication, removing the limitation on genome size; and, finally, the origin of cell organelles, in particular the mitochondrion and, in algae and plants, the plastid. Of these events, at least the last two qualify as major transitions in the sense of being major changes in the way genetic information is stored and transmitted.

Another feature emerging with the eukaryotes is the process of meiosis, leading to haploid cells and the merger of two such cells to form a new cell. In short, most of those features now recognized as sexual reproduction first occur in the protists. The emergence of sex has been extensively discussed; we here note that it parallels the emergence of the eukaryotes and provides a programmed method of combining genetic information. Like endosymbiosis, it provides a route to richness of variety in organisms.

The acquiring of endosymbiotic bacteria by eukaryotes has emerged as a method of complexification. In a number of taxa, bacteria live within various cells and may contribute to aspects of the metabolism. Protozoa living in the guts of termites are such an example. The protozoa have bacterial endosymbiants. The digestion of wood appears to depend on chemical activities of all three taxa. Emergent endosymbiosis that led to eukaryotes appears to be an ongoing feature of the present-day world.

## Chapter 13—Readings

Margulis, Lynn, 1998, *Symbiotic Planet,* Basic Books.

Margulis, Lynn, McKhann, Heather I., and Olendzenski, Lorraine, 1993, *Illustrated Glossary of Protoctista,* Jones and Bartlett.

Margulis, Lynn and Schwartz, Karlene, *Five Kingdoms,* 3rd Edition, 1998, W. H. Freeman and Company.

Smith, John Maynard, and Szathmary, Eörs, 1999, *The Origins of Life,* Oxford University Press.

# 14

# Multicellularity

The emergence of prokaryotes was a matter of macromolecular chemistry, and chemical networks developed with the emergence of metabolism. The emergence of eukaryotes developed from chimeras of prokaryotes sampling numerous ways of putting functional parts together. Three major foci of energy metabolism ultimately developed: autotrophy (largely photoautotrophy), ingestion of soluble molecules excreted into the environment by other organisms, and eating of other organisms. When these forms settled down, they were on the way to being plants, fungi, and animals, but still were unicellular or clusters of cells, i.e., colonial organisms. Another feature of the eukaryotes' emergence is the meiotic cycle, in which an ordinary cell undergoes two consecutive cell divisions leading to a haploid cell with half the number of chromosomes of a diploid cell. The combination of two haploid cells leads to a diploid cell. This is the normal method in which eukaryotes give rise to progeny.

The next emergence is multicellularity, in which an adult organism consists of many cells with different functions resulting in more complex forms. Since meiotic replication results in a single-cell stage of an organism, a subsequent series of cell divisions with differentiation must take place to produce the multicelled form of the organism, starting with the unicellular diploid form. This process is known as development or morphogenesis. The hereditary material must contain the program for the entire life cycle of the organism. To take an elaborate example, the butterfly genome must contain not only the specification of the butterfly, but also the specification of the caterpillar. In our subsequent search for emergences, we focus on the animalia, but it is important to realize that animals, plants, fungi, protists, and bacteria in an ecosystem all evolve simultaneously. All evolution is coevolution, since the species of an ecosystem are all interdependent.

Our interests, while embedded in this general biology, ultimately focus on the arrival of humans and human society. We will have little to say about plants, fungi, arthropods, mollusks, annelid worms, and other taxa. They are all there as part of the biological background within which mankind rises, flourishes, and tries to understand our universe and our planet.

In the evolution of the taxonomic tree, multicellularity has occurred many times. We shall review some of these instances and then focus on a particular kind of multicellularity that has led to the animals. Among single-cell organisms, it often happens that cells divide and do not split apart, but show a stickiness that leads to colonies. When this is followed by variable differentiation of the cells of a cluster into different forms, we refer to the phenomenon as morphogenesis.

Some simple cases exist such as cyanobacter that grow in chains, some cells being chlorophyll-containing and photosynthetic, with others differentiating into colorless nitrogen-fixing heterocysts. Actinobacteria grow in fungal-like filaments, some of which differentiate into spore producers. They are, however, prokaryotes. Brown algae or kelp are very large protoctists that may grow to ten meters or more in length. Because they do not produce embryos as the product of fusion, they are not classified as Plantae (plants), but are clearly multicellular. Similar statements may be made about the red algae and other multicellular forms.

Three major groups of true multicellular organisms emerged from the protoctista: the fungi, the plants, and the animals. Animals are multicellular organisms that usually develop from a blastula, a developing embryo formed from a diploid zygote, the result of the fertilization of a haploid sperm and haploid egg, followed by a few cell divisions. Fungi have chitinous cell walls and propagate by spores. Fungi may be single-celled as yeast or highly multicellular with specialized cell types such as the mushrooms. Plants have haploid and diploid stages of the life cycle. Diploid embryos are sometimes retained by the haploid organism during development. The three major taxa are distinguished by modes of nutrition. Plants make their nutrients photosynthetically. Fungi are saprophytes and absorb soluble molecules produced by other taxa, while animals ingest and digest plants, fungi, and other animals.

We narrow our focus at this point to the emergence of various forms of multicellularity among the animals. (It is no less interesting in other forms, but we are on our way to focus on the emergence of humans, who are clearly fauna.) Since animals at some stage develop from a single fertilized egg, development into a structured organism requires a morphogenetic process. This may take many forms. The smallest adult cell number occurs in the dicyemeds (Rhombozoa), which are tiny worm-like forms. They have a central cell surrounded by a layer of a species-fixed number of jacket cells, the order of 20. Within the central cell, new organisms develop from reproductive acroblast cells. This is the unusual case of cells growing within cells, or new organisms growing within cells. There is an elaborate life cycle. These unusual animals are parasites of benthic invertebrates such as the octopus. Organisms with a fixed number of somatic cells are called eutelic. Rhombozoa are the simplest of the many eutelic organisms and are composed of 20 to 30 somatic cells.

Eutely also occurs in multicellular animals with 500 to 1,000 cells. The phenomenon is found in many taxa, of which nematodes and rotifers are the best studied. Constancy of cell number requires a very tight morphogenetic programming quite different from that of most metazoans, which seem to specify development at the tissue level rather than the cell level. The development plan of the nematode *C. elegans* thus has been spelled out in greatest detail going from one cell to 982 cells.

Since cell dimensions seem to be constrained by such factors as diffusion rates that place limits in the few micron range, the emergence of multicellularity seems to be a prerequisite to the evolution of plants and animals—organisms that function at the macroscopic level. There are a few species of large (centimeter-range), single-celled organisms, but they seem to have developed special methods of transport. An example would be the green alga *Acetabulara mediterranea*. Understanding morphogenesis is a central area of research in modern biology. To some it is the central area.

Note that *Acetabularia* (among the largest of cells) has a volume of about 5 cubic centimeters, while mycoplasma (among the smallest of single cells) can have volumes of around $10^{-13}$ cubic centimeters. The ratio of cell sizes is a surprising 50 billion while cellularity is clearly the unique form of biological organization, the flexibility of the mode is quite extraordinary.

Another group of animals, the sponges or Porifera are large and made of two cell layers. They lack the usual organ structure of most animals and appear like a colony of cells with some differentiation. They do not seem to be on the main evolutionary pathway of the other animals, but the evolution of the earliest animals is very uncertain. This is largely due to the absence of fossil remains by organisms that had no shells, skeletons, or even cell walls. We must reconstruct the evolution from properties of extant taxa and, most recently, from macromolecular sequences of contemporary forms. This subject is somewhat in its infancy but should shortly give us much more detailed information.

In any case, multicellularity and differentiation are emergent properties. The agents are cells; indeed, a clone of cells derived from a single fertilized egg. The interactions and internal instructions lead to cell differentiation. The resultant structures are selected by fitness within environments and habitats.

An organism consists of a population of interacting differentiated cells. Since the cells are descended from independent unicellular organisms, there may be situations in which cells and organisms have conflicting goals. Cancer is the best known of these, where the growth and fitness of one cell line is detrimental to the organism and eventually to that cell line. Morphogenesis and differentiation are truly novel emergences in which the world of cells becomes the world of organisms. The relation between the organism and its cells has been discussed in detail in *The Evolution of Individuality* by Leo Buss.

A central theme in discussions of evolutionary biology is the extent to which complexity has increased over the four-billion-year history of change. In the absence of an agreed-upon metric of complexity, it is difficult to formulate a precise analysis. Nevertheless, a number of increasingly rich databases of relevant information allow us to revisit the question of evolutionary complexity in terms of some molecular, histological, and neurobiological approaches.

The first two billion years are the age of the prokaryotes, in which complexity differences are largely chemical, since there is a similarity in cell morphology including the absence of membrane-bounded organelles. Under these circumstances, genome size provides a measure of complexity, since it limits the number of enzymes and other functional macromolecules. If we assume that the earliest organisms were autotrophs, meaning that all carbon compounds are derived from one-carbon precursors, then the TIGR Microbial Database presents genome values of 1.5 and 1.75 megabases for two such species: *Aquifex aeolicus* and *Methanobacterium thermoautotrophicium*. These are significant limiting values, since autotrophs must possess all reactions essential to the core metabolic chart starting with some one-carbon components such as $CO_2$ or $CH_4$. The genome sizes can code about 1,200 functional proteins and RNAs. When we find the minimal genome of a chemoautotroph, we will have the minimal cell that could have been the universal ancestor.

The genome size can be reduced by herotrophy, or, in the extreme case, by parasitism or endosymbiosis. It can go up by adding appropriate chemical steps to be competitive in one or more niches. Clearly in the category of gene loss by parasitism are the mycoplasma, chlamydia, and rickettsia, with genome sizes ranging from .58 to 1.28 megabases. Most prokaryotes in the TIGR database range from 1.5 to 5.0 megabases, so that a certain general range of complexity within a factor of about ten describes this mode of cellularity.

In the usual evolutionary scheme, the prokaryotes gave rise to the protoctista, a large group of organisms that are unicellular or colonial-unicellular and tend not to have a tissue grade of development. The complexity of this group is characterized by a great variety of cell organelles: nuclei (often several per cell), granules, vacuoles, undulipodia, plastids, endosomes, oil vesicles, chloroplasts, paramylon bodies, kinetisomes, and others (*Five Kingdoms,* Margulis and Schwartz). Thus, although the group stays unicellular, complexity in cell structure allows these organisms to exist in almost all aqueous niches. There is insufficient data on genome sizes of organisms in this taxon to relate this parameter to complexity.

The protoctista gave rise to the three higher phyla: fungi, plants, and animals. The genome of the yeast *Saccharomyces cerevisiae* has been sequenced and has a genome size of 13 megabases. The entire kingdom of fungi seems to range between about 10 and 175 megabases, clearly larger than the prokaryotes. The largest reported bacterial genome, *Streptomyces coelicolor* (8.0 megabase), is not much smaller than the lower limit of fungal genomes.

Genome sizes among the animalia range from 30 to 21,000 megabases with no simple taxon-size relations. Thus, although genome size may serve as a complexity measure for the prokaryotes, it seems inappropriate for the animal kingdom and other higher eukaryotes.

In a fascinating paper entitled "Morphological Complexity Increase in Metazoans," by Valentine, Collins, and Meyer, the number of cell types is used as a measure of complexity. The use of cell type requires a certain agreement among histologists as to the classification of cells. This demands a standard "degree of lumping and splitting of cell types."

Table 3 shown below presents the cell-type number as a function of the number of years since the beginning of the Cambrian Age. Varieties of nerve cells are not included in the table. I have discussed this with neuroanatomist Ann Butler, who points out the great diversity of neural cell types and consequent difficulties in quantifying them, but suggests that the overall curve shape of cell type versus time would not be radically

TABLE 3: NUMBER OF CELL TYPES

| Years since beginning of Cambrian Age | Number of cell types |
| --- | --- |
| 0 | 2 |
| 70 million | 75 |
| 170 million | 125 |
| 270 million | 160 |
| 370 million | 180 |
| 470 million | 200 |
| 570 million (present time) | 220 |

altered by adding this number and arriving at the total number of cell types.

Two features characterize the cell type versus time relationship: (1) it is monotonically increasing, and (2) it appears to be approaching an asymptote. If the parameter chosen is a useful measure of animal complexity as suggested by the authors, then complexity as measured by varieties of cells has been systematically increasing over the entire epoch.

The paper suggests that different measures of complexity may be appropriate to different taxonomic radiations. Genome size seems appropriate to the prokaryotes, and number of cell types seems appropriate for the animalia. This is all clearly context-dependent.

## Chapter 14—Readings

Buss, Leo, 1992, *The Evolution of Individuality*, Princeton University Press.
Margulis, L., and Schwartz, K. V., 1988, *Five Kingdoms*, W. H. Freeman Co.
Valentine, J. W., Collins, A. G., and Meyer, C. P., 1994, "Morphological complexity increase in metazoans," *Paleobiology*, 20:131-142.

# 15

# The Neuron

The key to multicellularity is that a single fertilized egg formed by the fusion of two haploid gametes can give rise in the morphogenetic process to a variety of specialized cells. Thus, switches must be available to turn on and off genetic sequences to distinguish the cell morphology and biochemistry in different tissues. A hair cell is vastly different from a liver cell, which is vastly different from a nerve cell, yet they are all in the clonal progeny of a single fertilized ovum. The number of cell types appears to increase along the evolutionary pathway indicated in the previous chapter.

The descendants of the earliest animals are probably the placazoa and the porifera (sponges). Communication between cells in the protists and early animals is largely chemical. There are two methods: (1) A cell releases into the environment molecules that freely diffuse and may adsorb on the surface of a second cell or be transported across the second membrane to a receptor site; (2) gap junctions may exist between neighboring cells that permit intercellular transport of matter, including signaling molecules. Thus, signaling between cells is limited by diffusion, which is a slow process over large distances, as well as ultimately being concentration limited by dilution.

A nerve cell, on the other hand, receives a chemical signal at a given locus on the surface and converts it into an electrical signal, the action potential. In the electrical form, the signal moves rapidly along the axon and triggers chemical release at contacts with receptor sites of other cells. The axon may be thousands of cell diameters in length, so that a cell-to-cell signal may be sent rapidly over large distances.

Now, animals may be sessile, such as sponges, or very small, such as dicyemids, and not require nerve cells, but a large and responsive animal will require rapid signalling between remote parts. The emergence of the neuron as a cell type was a critical factor in animal evolution. It may have taken up to a billion years from the first multicellular animal to animals with nervous systems. The agents were cells, and the selection was for cells that could perform certain tasks.

The most primitive nervous system is in the Cnidaria or coelenterates, animals such as hydra or jellyfish. Their axons lack the myelin sheath found in other animals, and the nerves are capable of signal transmission in both directions. The nerve cells form a loose network throughout the organisms. The related phylum Ctenophora, or comb jellies, also have diffusion. For a number of reasons, it appears that nerve cells were derived from epithelial cells. In present-day animal embryos, epithelium and nerves derive from the same tissue layer, indicating a close relationship. The necessity for excitable cell membranes seems clear and can be traced back to protozoa, yeast, and bacteria.

In trying to track the oldest and simplest neural system, we find the following description by Professor Andrew N. Spencer:

> It was probably in an ancestral cnidarian that the earliest evolutionary experiments in neuro-neuronal and neuro-effector communication were played out. By studying present day cnidarians, we hope we are examining those synaptic mechanisms which were selected for, perhaps as far back as the Pre-Cambrian era, and which have been conserved with only minimal modification. Of course, we cannot be certain that physiological evolution proceeded at the same rate as morphological changes, nevertheless, the close resemblance of extant forms to fossilized imprints of cnidarians from this era (for example the Ediacara fauna of Australia) hint of slow rates of evolution.

Spencer then elaborates:

> ... many of the basic synaptic mechanisms and properties that we associate with more "advanced" nervous systems, such as Excitatory and Inhibitory Post-Synaptic Potentials and Miniature End Plate Potentials, facilitation, temporal and spatial summation, and $Ca^{++}$-dependent release of transmitter, can be demonstrated in the Cnidaria. With some danger of oversimplifying, one could say that it was in this phylum that most of the important properties of syn-

apses evolved, and that since that time, most evolutionary change in higher nervous systems (major protostome phyla and the chordates) has been with respect to the complexity of connections.

Next let us consider a typical nerve cell. It consists of a cell body containing the nucleus and much of the cellular apparatus, such as mitochondria. Extending out from this core are two types of structure, axons, which are usually quite long, and dendrites, which are much shorter than the axons. The axons terminate in synaptic knobs that may contact other neurons, muscles, or sensory organs. The axons of other nerve cells usually form synapses with the dendrites.

Signals are usually transferred chemically at the synapses between cells. This can trigger an action potential, sending an electric signal along the axon. The action potential travels along the axon by a depolarization of electric charge across the membrane. To prepare the nerve to fire again, ions must be pumped across the membrane to restore the potential. Nerve conduction is thus an active process requiring energy. Since there is a recovery time for restoring the potential, the rate at which nerves can fire is limited. At the terminus of the axon, the signal is once again transmitted chemically. The nerve cell, or neuron, is thus a switching station, transmission line, and computer. As a result, such cells are capable of a sophisticated repertoire of activities. Large networks of cells of this type can perform an arbitrarily complex set of computations and responses. A mammalian brain may have more than a trillion nerve cells.

Neurons of more advanced animal taxa tend to be: (1) motor neurons whose axon terminals connect to muscle, (2) sensory neurons that are connected to receptors, (3) interneurons between sensory and motor neurons, and (4) neurosecretory neurons. In primitive organisms such as coelenterates, the neurons tend to be multi-functional, rather than specialized in the way noted above. The neurotransmitters in these taxa are mostly neuropeptides. The basic features of the nerve cells have been quite conservative over the evolutionary epoch, but the structures of networks of neurons have changed radically from the hydra neural nets to the mammalian brain. Many of the more primitive neurons appear to be bidirectional.

In any case, the first neurological emergence was the neuron itself with all the potential of that cell type. The second emergence was the variety of specialized nerve cells. The third emergence was the elaborate network of various nerve cells that ultimately led to the brain.

The neuron was a landmark in animal evolution. It may have been one of the features that led to the Cambrian explosion of animal forms.

How the first neurons arose remains speculative. It is known that in many hydras epithelial cells can conduct action potentials. Some sponge cells exhibit action potentials. This suggests an evolutionary pathway.

In any case, the neuron is an emergent structure that arose about 700 million years ago. Once evolved, it rapidly acquired many of the features seen in more advanced neural systems. The existence of neural nets permitted the selection of fit behavior from the array of all possible behaviors. To some, this would be an emergent step on the way to mind, and I agree.

As a graduate student, I attended a lecture by the renowned photographer Roman Vishniac, who had turned his attention to the cinema-photography of freshwater-pond organisms. His film ran about 30 minutes, and I remember being awed by how purposeful the activities of many of these organisms seemed. I assume that most of the organisms were the protoctista that we have been discussing.

In our review of emergences, we are somewhere between molecules and minds, and lest minds come upon us too suddenly, let's consider some mental or noetic-seeming aspects of the emergences we have been discussing. We give three instances: the behavior of electrons, prokaryotes, and protoctists.

Recall that in our discussion of the Pauli exclusion principle we dealt with the restriction that no two electrons in a structure can share the same four quantum numbers—presumably four quantum numbers because of the four dimensions (three in space and one in time) In formulating the Schrödinger equation using relativistic quantum mechanics. This principle does not come from the dynamics of the problem, but from symmetry requirements on the solutions.

Note that the Pauli principle organizes all of physics and chemistry. Because of the nondynamical feature, several physicists and philosophers of science detect a kind of noetic feature deep in physics.

The level that we can first recognize as behavior occurs in the prokaryotes. Somewhere in bacterial evolution, motility appeared. The operative structures are flagella, which rotate, propelling the cells. A number of cases were discovered in which cells in a gradient of nutrient swim toward higher concentration, and in a gradient of toxins swim toward lower concentration. The mechanism is somewhat indirect. Periodically the swimming cells randomly switch directions. In a favorable gradient they change less frequently, and in an unfavorable gradient they change more frequently. They are letting their profits run and cutting their losses. For a population of cells, this leads to a fit behavioral repertoire.

The behavior looks causal, but the endpoint looks teleological. It requires sensing the environment, concentration versus time, and responding to the time gradient, which is also a space gradient, since the organisms are swimming. I think it important to look at these hints of cognitive behavior as they appear. It looks teleological because the organism must respond, whether the sensed concentration gradients are good or bad.

The next behavioral level emerged with the protoctists. In Roman Vishniac's film about pond-water organisms, this collection of mobile protozoans, euglenoids, and other swimming organisms pursued their lives as predators, grazers, and prey, exhibiting an awesome array of motions and behaviors. Here were largely single-celled creatures without nervous systems pursuing elaborate behavioral repertoires.

In 1912, the zoologist Jennings described the following experiment with the protozoan *Stentor*. Because the behavior seems to be purposeful as in the Vishniac film, I present his extended description:

> Let us now examine the behavior [of *Stentor*] under conditions which are harmless when acting for a short time, but which, when continued, do interfere with the normal functions. Such conditions may be produced by bringing a large quantity of fine particles, such as India ink or carmine, by means of capillary pipette, into the water currents which are carried to the disk of *Stentor.*
>
> Under these conditions the normal movements are at first not changed. The particles of carmine are taken into the pouch and into the mouth, whence they pass into the internal protoplasm. If the cloud of particles is very dense, or if it is accompanied by a slight chemical stimulus, as is usually the case with carmine grains, this behaviour lasts but a short time; then a definite reaction supervenes. The animal bends to one side. . . . It thus as a rule avoids the cloud of particles, unless the latter is very large. This simple method of reaction turns out to be more effective in getting rid of stimuli of all sorts than might be expected. If the reaction is not successful, it is usually repeated one or more times. . . .
>
> If the repeated turning toward one side does not relieve the animal, so that the particles of carmine continue to come in a dense cloud, another reaction is tried. The ciliary movement is suddenly reversed in direction, so that the particles against the disk and in

the pouch are thrown off. The water current is driven away from the disc instead of toward it. This lasts but an instant, then the current is continued in the usual way. If the particles continue to come, the reversal is repeated two or three times in rapid succession. If this fails to relieve the organism, the next reaction—contraction—usually supervenes.

Sometimes the reversal of the current takes place before the turning away described first; but usually the two reactions are tried in the order we have given.

If the *Stentor* does not get rid of the stimulation in either of the ways just described, it contracts into its tube. In this way it of course escapes the stimulation completely, but at the expense of suspending its activity and losing all opportunity to obtain food. The animal usually remains in the tube about half a minute, then extends. When its body has reached about two-thirds its original length, the ciliary disc begins to unfold and the cilia to act, causing currents of water to reach the disc, as before.

We have now reached a specially interesting point in the experiment. Suppose that the water currents again bring the carmine grains. The stimulus and all the external conditions are the same as they were at the beginning. Will the *Stentor* behave as it did in the beginning? Will it at first not react, then bend to one side, then reverse the current, then contract, passing anew through the whole series of reactions? Or shall we find that it has become changed by the experiences it has passed through, so that it will now contract again into its tube as soon as stimulated?

We find the latter to be the case. As soon as the carmine again reaches its disc, it at once contracts again. This may be repeated many times, as often as the particles come to the disc, for ten or fifteen minutes. Now the animal after each contraction stays a little longer in the tube than it did at first. Finally it ceases to extend, but contracts repeatedly and violently while still enclosed in its tube. In this way the attachment of its foot to the object on which it is situated is broken and the animal is free. Now it leaves its tube and swims away. In leaving the tube it may swim forward out of the anterior end of the tube; but if this brings it into the region of the cloud of carmine, it often forces its way backwards through the substance of the tube, and thus gains the outside. Here it swims away, to form a new tube elsewhere.

... the changes in behaviour may be summed up as follows:

(1) No reaction at first; the organism continues its normal activities for a time.

(2) Then a slight reaction by turning into a new position.

(3) . . . a momentary reversal of the ciliary current . . .

(4) . . . the animal breaks off its normal activity completely by contracting strongly . . .

(5) . . . it abandons its tube . . .

There are a wide variety of understandings of what is meant by mind. The reductionist behaviorist tradition would argue that mind is an epiphenomenon of the activities of collections of neurons. They argue that minds do not in fact exist. At the opposite extreme, the idealist tradition going back to George Berkeley would argue that mind is all that exists, and matter is an epiphenomenon posited by minds for explanatory purposes.

The Kantians would argue for the existence of both mind and matter, the latter being the *ding an sich* (thing in itself) that minds aspire to and cannot fully comprehend.

Our view is that all of the above is too simplistic. The universe, whatever its ultimate character, unfolds in a large number of emergences, all of which must be considered. The pruning rules of the emergences may go beyond the purely dynamic and exhibit a noetic character. It ultimately evolves into the mind, not as something that suddenly appears, but as a maturing character of an aging universe. This is something that we are just beginning to understand and, frustrating as it may be to admit such a degree of ignorance, we move ahead. That is our task as humans; some would call it knowing the mind of God and regard it as a vocation.

There are those who would argue against the appearance of anything mental at a primitive level. Gregory Bateson, in his book *Mind and Nature,* argues:

> In this matter, I prefer to follow Lamarck, who, in setting up postulates for a science of comparative psychology, laid down the rule that no mental function shall be ascribed to an organism for which the complexity of the nervous system of the organism is insufficient.

The Bateson-Lamarck viewpoint assumes that the "mental" is a property of animals with an elaborate nervous system. While many scientists agree with this approach, I would argue that noetic features emerge earlier in the evolution of our planet. Mind emerges over a long time, not just over the last 500 million years. It is more deeply embedded in an evolving universe and may have prebiotic roots. In the next chapter I shall move to the consequence of the emergence of nerve cells, a necessary step for higher mental activity.


## Chapter 15—Readings

Bateson, Gregory, 1979, *Mind and Nature: A Necessary Unity,* Dutton.

Briedbach, O., and Kutsch, W., 1995, *The Nervous System of Invertebrates,* Birkhäuser Verlag.

Griffin, Donald R., 1992, *Animal Minds,* University of Chicago Press.

Jennings, H. S., 1906, *Behavior of Lower Organisms* (1967 reprint), Indiana University Press.

Spencer, Andrew, 1989, in *Evolution of the First Nervous Systems,* Peter A. V. Anderson, Editor, Kluwer Academic Press.

# 16

# Animalness

The early phase of multicellularity probably started with a somewhat colonial-type organism such as the sponge (Porifera). Even these relatively simple-looking organisms have several cell types and a life cycle including development of sperm and eggs, fertilization, the development of motile larval forms, and finally the attachment of the larvae to surfaces followed by the growth of adult sponges. Strange as it may seem, organisms like sponges were probably our earliest multicellular animal ancestors.

The next major emergence in animal evolution is the nerve, discussed in the last chapter, a cell type that permits behavior, a primary feature of this major taxon. With the appearance of neurons, the animalia were on the evolutionary path from sponges to humans.

Reconstructing the first hundred million years of animal evolution has proven to be extremely difficult from a technical perspective. The early animals were soft-bodied creatures like jellyfish and worms, which left few or no fossils. We thus have been required to construct the evolutionary tree from the morphology and physiology of present-day descendants of primitive animals that have undergone as much as 700 million years of evolution from their earliest ancestors. The taxonomy based on nucleic acid sequence is a related subject of present-day investigation, but at the moment lacks sufficient data to be definitive and also deals with present-day organisms. It seems reasonable to many evolutionary zoologists to assume that the first organized multicellular animals were coelenterates something like our present-day hydra, corals, and jellyfish. These organisms are highly multicellular, have a number of cell types including neurons, and are morphologically programmed at the tissue level. There is clearly a complex evolutionary pathway from the ancestral protoctist to the coelenterates, but little can be said of the intermediates or the time it took for the various transitions. The simplest and presumably most primitive extant animals are Placozoa. These two- or possibly three-layered amebalike creatures have a dorsal/ventral (back/front) axis but no indication of anterior/posterior (head/tail) or left/right directionality. There seem to be no neurons in these animals. The sponges (Porifera) are not motile in most stages of the life cycle, but are sessile. They pump water, filter it, and trap suspended food particles. They also lack neurons, but appear to have conducting epithelial cells from which electrical signals have been recorded.

New features arise with the appearance of the coelenterates (Phylum Cnidara). Examples are hydra, jellyfish, and corals. While the placazoans and sponges are colonial animals, the coelenterates have a considerably more organized body plan with a radial symmetry that requires two spatial specifications along the cylindrical axis and the radial axis in order to specify the morphology. Some species have a bilateral symmetry imposed over the radial symmetry. This points to subsequent emergent features of animalness.

The coelenterates are probably the first phylum to have a nervous system, a network of neurons, connected by synapses and going to most parts of the organism's body. Most species have a larval stage although there are a number of life cycles observed in different species.

Full bilateral symmetry is found in the flatworms (Phylum Platyhelminthes). The free-living forms have paired eye spots in the head end, connected by a brain that is attached to lateral ganglia parallel to the axis of the worm. The mouth is on the bottom side toward the central rear portion of the animal.

The early animal evolutionary sequence is: Protoctists (perhaps Zoomastigota), Placozoa, Coelenterates, Flatworms (Turbulara or Planaria). The basic emerging features of animalness are sensory organs, a nervous system, and a digestive tract to trap and ingest or digest (or digest and ingest) other organisms or other particulate food.

Two features emerged in the first hundred million years of animal evolution, and we are unsure of the order or independence of emergence. These features are:

1. The transition from radial symmetry to bilateral symmetry at some stage of the life cycle. The morphogenetic program of cylindrically symmetrical animals has two dimensions, radial and axial. True bi-

laterally symmetrical animals require left/right, anterior/posterior, and dorsal/ventral specifications. However, bilaterally symmetrical animals without the dorsal/ventral distinction require only two dimensions of specifications and could be related to both coelenterates and the flatworms (Platyhelminths).

2. Cephalization, the collecting of sensory and neural elements at the anterior end of the organism. This is the beginning of an evolutionary pathway that culminates in a brain in the head end. There is an axial organization and a direction of motion. Thus there is a tendency to put sensory function at the front end in the direction of motion, an evolutionary development of obvious value.

In the evolutionary tree of the animals, a major branch point occurs with the flatworms or some near relative of the free-living tubelaria. To understand the split, we start with the definition of animals as set forth by Margulis and Schwartz. Animals may be defined, according to these authors, as "heterotrophic, diploid, multicellular organisms that usually (except sponges) develop from a blastula. The blastula, a multicellular embryo that develops from a diploid zygote produced by fertilization of a large haploid egg by a smaller haploid sperm, is unique to animals." The next taxonomic division occurs because of events at the blastula stage. In one group, the first opening in the blastula, the blastopore, becomes the mouth. These animals are the protostomia and include annelid worms, arthropods, and mollusks. In the other group, the opening becomes the anus, and a second opening becomes the mouth. This group includes echinoderms, arrow worms, hemichordates, and chordates. The split into these two super-taxa occurred about 600 million years ago. Surprisingly, genes for eyeness, segmentation, and nerve and brain organization have persisted from the original higher animal ancestor into both major branches of the animal tree. Fruit flies and mice have genes in common.

This divergence is clearly very important, but it is difficult to find a scenario that accounts for the two developmental patterns or indicates which would be most fit under what conditions. Presumably a better understanding of morphogenesis will make this matter clearer.

From an ecological point of view, animals are either grazers who eat photosynthesizers, carnivores who eat other animals, or specialized saprophytes who eat solubilized products of photosynthesizers and other animals. The emergence of animals requires ecosystems that provide the necessary nutrients. Plants and fungi get their food rather passively, while animals often seek their nutrients actively. This explains why behavior is

TABLE 4: EMERGENCE OF ANIMALS

Millions of years ago

| 20 | Hominids |
|---|---|
| 140 | Main groups of mammals |
| 160 | Birds |
| 240 | Dinosaurs, stem mammals |
| 300 | Mammal-like reptiles |
| 340 | Primitive reptiles |
| 360 | Amphibians |
| 380 | Insects |
| 400 | Lungfish and crosspterygians |
| 420 | Fish with jaws, bony fish |
| 440 | Sharks |
| 480 | First jawless fishes |
| 520 | First animals with a notochord |

a primary feature of animals among the multicellular taxa. They are either pursuing food or fleeing from pursuers. Behaviors for acquiring nutrients represent major fitness criteria for animals and are components of all evolutionary emergences among animal taxa.

At this stage we are launched into the very complex and endlessly fascinating subject of animal evolution. We examine only a few of the many, many emergences from hydra to humans. We also set aside the plants and fungi, ever mindful that the evolution and origin of species always takes place in an ecosystem with a broad range of different taxa. Truly no species is an ecosystem itself, and a fuller treatment would include coevolution, the evolution of all species in an ecosystem, as well as symbiosis in all its forms.

Table 4 sets out a time line, of uncertain accuracy, but sufficient to see in some perspective the chronology of major emergences that constitute animal evolution. The first 120 million years of chordate evolution is a study of life in the sea. This is followed by amphibians, reptiles, and mammals.

The timetable of life has constantly sped up in degree of complexity over its four-billion-year lifespan. The first two billion years were restricted to prokaryotes. The next billion years saw the rise and spread of protists. In the next half-billion years came multicellularity, and the Cambrian explosion of the major animal taxa followed.

A quarter of a billion years more led to the first mammals and the age of dinosaurs. With the disappearance of the dinosaurs 65 million years ago, a rapid radiation of the mammals and birds led to the rich variety of Pleistocene animals. The extensive evolution leading to the 100-ton blue whale and the 2-gram Etruscan shrew all took place over only those 65 million years. Primate evolution has proceeded even faster, leading ultimately to humans, where cultural evolution has intervened, speeding up evolution by several orders of magnitude.

Sometime around one billion years ago, the first animals emerged from protoctist predecessors. Animalness has persisted and developed great complexity, demonstrating modes of morphogenesis in which prior advances have been retained. In spite of the great elaboration of form, animalness itself is sufficiently defined that it can be recognized structurally and morphogenetically over the millions of species that have evolved from the universal animal ancestor.

## Chapter 16—Readings

Margulis, Lynn, and Schwartz, Karlene, 1998, *The Five Kingdoms,* W.H. Freeman and Co.

# 17

# Chordateness

We next focus on one of the branches of the protostomia/deuterostomia divide. The figure shown below (p. 113) is the dendrogram or branch of the animal tree of the deuterostomia. This is the branch that includes us. As with all such dendrograms, there is some uncertainty in where they are rooted. That is, the original flatworm may have branched off to the left of the echinoderms or may have given rise to an earlier, no longer extant taxa that gave rise to the echinoderms and a second branch of all the Chordata.

The echinoderms include starfish, sea urchins, and sea cucumbers. The adult forms tend to have five-part or radial symmetry, but the larvae are bilaterally symmetrical, thus relating to their predecessors (the flatworms) and successors (the cephalochordates). It is not certain that echinoderms are on the main evolutionary pathway to chordates, but for convenience we adopt this view.

Early deuterostomes are Chaetognatha and Hemichordata. The former are axially organized animals that release neurotoxins at the head end, thus paralyzing prey that are then ingested. The Hemichordata are also axial in structure and have a primitive heart vesicle to help in propelling blood. They have gill slits and a proboscis and mouth near the head end. The general subsequent evolutionary pathway seems to be from the hemichordates to the chordates.

The tunicates are formally part of the group Urochordata, which includes sea squirts and ascidians. The larvae are free-swimming and bilaterally symmetrical, and the adult is sessile, pumping water through the animal and collecting food suspended in the water. In many earlier animals, the larval forms are very different from the adult, varying in anatomy and physiology. In the tunicates the notochord occurs in the tiny larva and innervates the tail, which makes up ⅔ of the structure. Emergent in Urochordata evolution are internal organs such as heart, liver, and stomach. Thus, features that ultimately become our innards emerged hundreds of millions of years ago. The specialized tissues of our evolutionary precursors eventually become well-defined organs to carry out special functions. The next group of taxa lengthens out: they are the Cephalochordata. (The evolutionary successors to the tunicates are, as just noted, the cephalochordates, or lancets, of which amphioxus is a surviving example often shown in biology books.) They show a head with no bony skull, but a collecting together of neural tissue, the first hint of a brain. They are wormlike with a head and tail end.

The adult cephalochordates have a resemblance to the larval tunicates. This process of retaining juvenile features into a new adult phase is called paedomorphosis. Thus the tunicates could have given rise to the lancets, small eel-like animals with a notochord with somites. They have gill slits, fin-rays, and a dorsal fin. They are intermediate between larval tunicates and primitive fish. In both tunicates and lancets the nerve chord terminates in a primitive brain. The larval forms are planktonic, and as they mature they settle down to burrow in the bottom, where they are filter feeders. The notochord is present in both larvae and adults.

Cephalization, the anterior collection of neural tissue in the head, occurred in both protostomes and deuterostomes. In the first case it eventually led to structures such as the cephalopod brain. Among the deuterostomes there are several invertebrate taxa leading to the emergence of chordateness characterized by a dorsal tube, the notochord, and gill clefts at some stage in the organism's life cycle. The notochord, according to Margulis and Schwartz, is "a long elastic rod that serves as an internal skeleton in the embryo of chordates replaced by the vertebral column in most adult chordates." The gill, according to the same authors, is a respiratory organ used for uptake of oxygen and release of carbon dioxide and regulation of diffusible ions by aquatic animals. Although we cannot state with certainty the chordate evolutionary pathway, the evolution of deuterostomes leading to animals like the tunicates is characterized by the development of a mouth and anus during morphogenesis. That is, they are characterized by an open alimentary canal. These animals universally have a blood supply and a method of oxygenation. They also have a coelum, a body cavity between layers of mesoderm, the median tissue layer in the developing embryo.

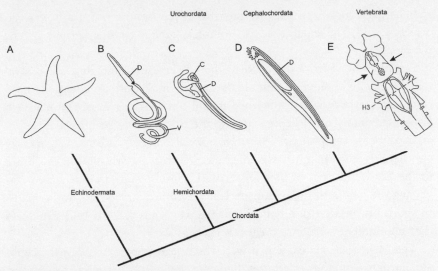

The Emergence of the Chordate Brain
Dendogram of the Deuterostomia leading to the craniate brain represented by a
cross-section of the lamprey brain. From Ann E. Butler *Chordate Evolution and
The Origin of the Craniates*

With the appearance of the first fish, the brain is protected by a cranium
made of cartilage as the structural skeleton. This group of primitive fish,
known as Agnatha, includes the present-day lampreys and hagfish. The
notochord has become the brain and spinal chord, and the general
vertebrate-like characters and body plan have emerged. The cartilage is
later replaced by bone and segmentation, giving rise to the spine.

The early evolution of chordates follows a pathway from marine worms
to primitive fish, lampreys, and hagfish.

Along the pathways of animal evolution, a new feature has emerged
that we have previously noted. Morphogenesis is at first envisioned as the
transformation from a fertilized egg to an adult. Often, as we have stated,
there are multiple stages, including free-living larval stages. Common examples of this phenomenon are the larval caterpillar and the mature butterfly, among the protostomia, and the larval tadpole and mature frog,
among the deuterostomia. From the point of view of coding, both morphological forms must be genetically encoded.

In discussing the evolution of a taxon, the entire time sequence of forms
and functions in a life cycle must be included. Thus, although tunicates
are presumably intermediate between flatworms and agnaths, the larval
tunicate more closely resembles the flatworm and adult agnath. The adult
tunicate is a sessile feeder that resembles neither. The diagram indicates

the changes from flatworms to agnaths. Although the transition was relatively rapid, the emergence of complex organs was a feature of the process.

A nerve chord from the head to the tail emerges, and this is a central feature of all subsequent animals. It is as if there were an emergent axial organizing principle that governed morphogenesis in this phylum. It also appears in many other groups among the protostomia such as the insects.

In any case, amphioxus (*Branchiostomia*) emerges as a paradigm early chordate. It is an inch or two long, axial in form, with mouth and brain in the head end. It is a filter feeder with gill slits and an elastic rod traversing the length of the organism. It has a circulatory system and gonads. In short, our relative of a half-billion years ago has features and structures that ultimately evolve into creatures such as we.

The earliest craniates, which were the intermediates between the cephalochordates and the vertebrates, had enlarged brains, paired sense-organs, and a head skeleton. It is possible to map the brains and brainlike structures across the taxa. Indeed, homologous and analogous structures develop in the protostomes. Lurking in the emergence of animals is a little-understood property of brain-ness.

## Chapter 17—Readings

Butler, Ann E., 2000, *Chordate Evolution and the Origin of Craniates, The Anatomical Record* 261, 111–125.
Margulis, Lynn, and Schwartz, Karlene, 1998, *Five Kingdoms*, W. H. Freeman and Co.

# 18

# Vertebrates

Up until the emergence of insects and amphibians, animal life was almost entirely marine. Within the vertebrates there was great evolutionary radiation, and a wide variety of fish evolved. These included:

- Agnatha (jawless fish)
- Sharks (cartilaginous fish)
- Bony fish
- Lungfish
- Crossopterygia

Thus, going from the lancets to the complex fish (that we treat as a single emergence) can easily be envisioned as a substantial number of emergences. Counting emergences can involve coarse graining or fine graining and can give almost any number we choose. There is something arbitrary in this book's present choice of 28 emergences. An effort is made here to deal with the major changes, but that is somewhat in the eye of the beholder. Over the entire evolutionary sequence of fish, the basic vertebrate body plan finally emerges: the skeleton, the major organs, and the brain with its major components. The sensory systems such as the eyes, nostrils, and lateral-line system to detect motion are all in place in our marine ancestors.

Because the sea bottom lacks oxygen and is a favorable environment for fossil formation, we have a remarkably detailed paleontological knowledge of 100 million years of fish evolution after the appearance of the first bony structures. In many ways, the known emergence of fish is the story of bone development. Firm structures are necessary to maintain shapes that are hydrodynamically efficient for moving through the water. The earliest such frameworks are made of collagen, a somewhat rigid protein, but after the biochemistry of bone formation evolved, bone (because of a number of its physical properties) became the method of choice for shaping vertebrates. Because all of the living world is generated from the same basic organic chemistry, it is often necessary to develop long and complex chemical pathways before the necessary molecules and the right structure are available for some macroscopic task.

We thus have a wealth of structural information, but little environmental and behavioral information about the evolution of the fish. The habitats and niches that fish occupied are difficult to reconstruct in the fluid oceans and rivers some hundreds of millions of years ago. Competition and fitness must have been the dominant selection factors, as in other evolutionary situations, but little else can be detailed of the behavior of these organisms. It leads us to ponder why fish require the development of such a complex brain.

The 150 million years from the emergence of fish, 520 million years ago to the first amphibians, 370 million years ago, witnessed the change from lancet-like creatures to the very sophisticated fish, crossopterygia. It was the age of the laying down of the vertebrate body plain: the bony skeleton, as well as the organs of physiological function coevolved. The earliest agnaths must have possessed a central nervous system, heart, liver, digestive tract, kidney, and circulatory systems. All subsequent vertebrate evolution builds on these structures. "In many ways, bone and external hard structures were the key to vertebrate evolution."

Steadman's *Medical Dictionary* defines bone as, "A hard connective tissue consisting of cells embedded in a matrix of mineralized ground substance and collagen fibers." The fibers are impregnated with calcium phosphate as well as a collection of other minerals that impart the hardness and strength. By weight, bone is composed of 75 percent inorganic material and 25 percent organic material.

During the period under discussion, most of the internal bodily structures that led to amphibians, reptiles, birds, and mammals emerged. It is frustrating that, other than the usual evolutionary explanations, we lack a more detailed theory of this emergence, since it established the structure and functions that led to us.

We can, however, suggest some selection criteria. Since life in the sea involves a search for food and escape from predators, the torpedo shape must be hydrodynamically favored and is hence characteristic of the group and is the basis of axial organization. Locating the sensory organs so as to receive signals from the direction of travel is obviously good engineering. Locating the mouth near the sense organs also aids in efficient feeding. Swimming toward food and away from fecal material organizes the gastrointestinal axis along the hydrodynamic axis. The vertebrate body plan is hardly an accident. It is primed by good engineering logic for life in the water. Choices appear in evolution. They are not random: they are constrained by physics and chemistry and would reappear, I maintain, in broad outline if the tape were replayed.

It is not a unique logic. Cephalopods and arthropods have solved the problems of feeding and escape differently, but there are a limited number of solutions, and axial organization with internal skeletons is one of the most efficient of these modes of operation.

The earliest fish, evolved from lancet-like animals, were jawless and had external shields made of bonelike material. There was a great evolutionary radiation of those jawless fish, and then, quite enigmatically, they became extinct, leaving only lampreys and hagfish, which are so specialized in their feeding habits and life styles that they can hardly be typical of this large order. These jawless fish must have been outcompeted by their evolutionary descendents, while the survivors found a niche as parasites in the world of the newly evolved bony fish.

The agnaths apparently gave rise to two major taxa, the sharks and their relatives, and the true bony fish. The sharks belong to the class Chondricthyes, meaning cartilage fish. We will not detail their emergence, as they probably are not on the evolutionary trail leading to the higher vertebrates, although again we note that all evolution is coevolution.

About 410 million years ago, a group of bony fish (Osteichthes) arose, and, as John A. Long notes in *The Rise of Fishes,* "Buried within the early evolution of osteichthyan fishes lies the key to the most complex transition in vertebrate history; how a water-breathing fish became a land-living amphibian." One of the emergent structures was a swim bladder, an organ of buoyancy, which eventually gave rise to the lung. Again, note that the sequences seem to go from tunicates to lancets to proto-vertebrates to the first fish to lungfish. The early ray-finned fishes had a bony vertebrate, a brain case, gills, fins, and a variety of internal organs not knowable from fossil evidence. The basic vertebrate pattern of a digestive system, circulatory system, and respiratory system had been well established by this point.

The emergence of a lung from the flotation organ was first seen in the

Dipnoi, or lungfish. There appear to be at least two separate emergences of lung breathing, so the lungfish are not considered ancestral to the amphibians. That role is believed to belong to the crossopterygia, or lobed-fin fishes. We are fortunate to have a living member of the group, coelacanths of the genus *Latimeria*. This group was thought to have been extinct for 50 million years, so the discovery of living representatives in 1938 was a high point in experimental ichthyology. Final emergent features of the crossopterygia are two sets of paired fins, pectoral and pelvic. This feature was clearly enabling when the fish began to invade the land and ultimately gave rise to the tetrapods. One need not be teleological, as the pelvic fins also aided in swimming.

Along with the sophisticated development of the vertebrate body plan during the age of fish, an equally surprising development of the fish brain occurred. Starting with the collection of nerve structures in the amphioxus head, a highly structured forebrain, midbrain, and hindbrain evolved in the emergent fish. Much of this accompanies the development of eyes and processing of visual information. Another portion dealt with the processing of olfactory information. The fitness advantages of a large and complex brain clearly show up in fish evolution. The ability to receive and represent sensory information must be an important feature of survival.

Over the long evolution from protists to higher fish, genes for morphogenesis emerge, which can be traced back to the sponges and appear in all branches of animal radiation. Certain genes seem associated with eyeness, others with segmentation. Cell-surface materials (glycoproteins and lipoproteins), by appearing in different forms and patches, led to morphogenic forms in biology in the same sense that bond angles and bond types lead to different forms of crystals in chemistry.

Once again we encounter emergences. The agents are cells with surface information (molecular stickiness). The processes are growth and sticking. Morphogenesis in all its forms emerges. This can be seen in what looks like crystal growth in sponges, to the exact cell-by-cell specificity in nematodes, to the elaborate morphogenetic organ pattern in fish, to the exquisitely elaborate networks of neurons in the human brain. This patterning can either be the result of force fields, concentration gradients, or surface forces. The genes can apparently drive a higher-order set of emergent structures. The process from genes to morphogenesis will be a major area of advance in the next decade of biological research.

It is strange to muse on the facts that our basic internal body and brain plans were formed deep in the ocean. Life inexorably moves into all the geospheres, and the biosphere covers the planet. That too is a feature of immanence and emergence.

## Chapter 18—Readings

Butler, Ann, and Hodos, W., 1996, *Comparative Vertebrate Neuroanatomy,* Wiley-Liss Inc.

Long, John A., 1995, *The Rise of Fishes,* Johns Hopkins University Press, Baltimore.

## 19

# Crossing the Geospheres: From Fish to Amphibians

The next emergence is in the traditional sense of the word as well as in its use in terms of complexity: "**emerge 1.** To rise up or come forth from or as if from immersion." In the evolutionary unfolding, one branch of vertebrates moved from water to land. All of our previous discussion has dealt with animals and other organisms living in water. The transition to land had ecological and physiological aspects. The planetary surface has been largely covered by water since early in Earth's history. When land appeared, living organisms found ways of exploiting the habitat and finding a home or better yet a niche. This occurred in all the major taxa.

The task was daunting. Since all organisms are from 50 percent to over 90 percent water, life on land requires mechanisms for getting and keeping adequate quantities of water. For plants, this is achieved by living on the shores of lakes and rivers or by putting down roots to acquire the water held in the soil. For animals, the usual methods are living in water, drinking water, or eating water-laden plants and animals. (There are more subtle methods, such as the camel's combusting fat and harvesting the $H_2O$ formed from the hydrogen in the lipid and the oxygen in the air.) Water must not only be found, but must be kept from evaporating and otherwise dissipating. This involves various skins and membranes of low water permeability.

To comprehend the inevitability of fish invading the land, let us return to the concept of a niche discussed earlier. After land became inhabited by plants and a number of invertebrates, including worms, crustaceans, and insects, a rich food source was available for a variety of aggressive carnivorous, herbivorous, and omnivorous fish. It was a food source just beyond their reach, but evolution within their own habitat had produced structures that could be modified for living out of water. A tetrapod-like crossopterygian fish and a primitive amphibian (Ichthyostegia) look remarkably similar. The relation seems clear. This need not be teleological. One can argue that among the vast variety of fish, some were best fitted for the invasion of land when the opportunity presented itself.

Respiration is the first problem. All higher animals are aerobes and require oxygen to combust food molecules, releasing the energy required for all living processes. In vertebrates the oxygen is carried in the blood and must diffuse across a membrane into the blood stream. In fish this occurs in the gills, as the oxygen is dissolved in the water in which the animals live. In land-dwelling animals the oxygen comes from the air, dissolves in the fluid in the lung sacs, and diffuses from there into the blood.

The lungs are not derived from the gills, but from an internal hydrostatic organ, the swim bladder, which was originally designed to help the fish regulate its vertical position in a water column. As this gas-filled organ folded and became vascularized, the swim bladder became a lung. There were fish that could gain oxygen both through the air bladder and the gills. The fish apparently began air breathing as a supplementary mode of oxygenation before moving onto land.

A second necessity for moving onto land was a mode of locomotion. This was anticipated by the paired limbs. The fins had many features of legs; the tetrapod pattern emerged before the transition to land. The first tetrapods or amphibians were not very different from their fish ancestors. In the larval stage they were still more closely related. All of this is decidedly teleological and would be rejected by most biologists, but for whatever reason the four fins were there when migration to land began.

Note that, from fossil and contemporary evidence, the evolution of lungs for air breathing seems to have occurred at least twice in the fish: once in the ancestors to the present-day lung fish, and once in the Crossopterygia. This suggests a certain inevitability or very high probability of vertebrates crossing the barrier from water to land. Again one thinks about what would have happened if the film were played over again.

Tetrapods, spending time both on land and in water, continually had to face the problem of dehydrating. For a long time tetrapods solved this problem by spending the most vulnerable periods of the life cycle in water;

thus, the eggs are deposited and fertilized in the water. The larval forms are fishlike and also live in the water. The animals are truly amphibians living in two geospheres, hydrosphere and atmosphere lithosphere.

Fertilization among the earliest amphibians occurred by the females laying eggs in water, subsequently or simultaneously accompanied by males depositing sperm. This necessitated early morphogenesis taking place in marine habitats. It rendered the embryos and hatchlings protection from dehydration but sensitive to predation. Among some modern amphibians, salamanders, and caecilians, fertilization is internal, allowing embryogenesis to take place in habitats in which the eggs are deposited. This is a step toward the reptilian pattern in which internal fertilization must take place because the eggs are shelled, and fertilization must precede the maturation of eggs.

An animal capable of spending an appreciable part of its life on land has emerged. As evolutionary biologist John A. Long has noted, "A small step for fishkind, but a great step for man." For terrestrial vertebrates, the transition begins the exploration of a whole new domain.

The earliest amphibians were the fishlike Ichthystego about 370 million years ago. The present-day members of this class are the burrowing wormlike caecilians, the jumping frogs and toads, and the salamanders that locomote on four legs. The present surviving taxa are probably quite unlike the amphibians that gave rise to the reptiles. They do however show how amphibians have come to occupy those niches between water and land.

Many amphibians are now among endangered and disappearing species. Living in three geospheres, they are particularly sensitive to habitat destruction, industrial toxins, and loss of suitable niches. In a sense, amphibians have been threatened ever since the emergence of their successors, the reptiles, about 280 million years ago. Most orders are extinct and known only through fossils. Within 100 million years of the emergence of the reptiles, nine orders of amphibians became extinct. Three orders were extinct even before the arrival of the reptiles. In a sense, this might be expected of a geosphere-crossing taxon. In the water they were not fully competitive with the fish, and on land they were not fully competitive with the reptiles and other successors. They are the ancestors of all terrestrial vertebrates including us.

The availability of organisms on land created a new niche for vertebrates able to take advantage of this food source. The occupation of this niche resulted in the emergence of amphibians. Both wormlike and tetrapod forms have evolved. Selection factors involved coevolution with insects and land plants. Given the geospheres, it seems clear that transition forms would evolve. Emergence of more fit terrestrial forms also seems to have been inevitable.

## Chapter 19—Readings

Cloudsley-Thompson, J. L., 1999, *The Diversity of Amphibians and Reptiles,* Springer.

## 20

# Reptiles

In the discussion of amphibians, we noted the transition from water-dwelling to land-dwelling taxa. In almost every major extant group there are organisms that spend their time on the lithosphere, in the hydrosphere, and in the atmosphere. Thus, while reptiles arose from amphibians as an adaptation to living on land, various groups such as the crocodilians have evolved to be primarily water-dwellers. So also with the mammals: the hippopotamuses have evolved to be partially aquatic, and the cetaceans, originally terrestrial, are almost entirely aquatic.

The reptiles have evolved from a branch of the amphibians. They breathe air, are cold-blooded, and have a covering of scales. The ova are fertilized internally, and a shell that also encloses food for the growing embryo covers the developing zygote. As a result most reptiles hatch as nominal adults, as distinguished from their predecessors that go through a larval stage, which must metamorphose into the adult form.

Although fish show sexual dimorphism between the females (egg-producers) and males (sperm-producers), both sexes release their sex cells in the water where fertilization takes place. The same process of fertilization is shown in many amphibians. In both cases behavioral patterns guarantee that the sperm and egg are laid down in the same place.

With the reptiles, something radically new emerges—the shelled egg, which contains the zygote and food supply encased in a shell, which is rather impermeable to water. Internal fertilization imposes an even stronger dimorphism, both anatomically and behaviorally. First, structures are required to transfer sperm to the eggs. This is at present done by reptiles and mammals with a penis, a vaginal cavity, and a method of insertion. Internal fertilization has developed independently among the protostomia, especially in several taxa of insects, and cephalopods.

This transfer of sperm can be accomplished by alternative procedures. Leaving the vertebrates for a moment and focusing on the mollusks, we note that in the male octopus one of the eight arms is specially adapted to reach into the sperm sac, pick up sperm cells, and place them in the egg sac of the female. An entire behavioral repertoire must be developed, as octopuses are normally very solitary animals.

Among a wide variety of taxa, mating among individual males and females developed as a method of securing internal fertilization. The end was to produce eggs that could develop and hatch, protected by a shell that prevented drying out. Achieving copulation and fertilization among animals that may be solitary and hostile became a biological imperative.

The receptivity of a given female and male to each other is of special significance. Among the vertebrates it imposed a different kind of selectivity on the evolutionary process. Fitness was no longer just survival, but locating a mate in order to transmit one's genes. Among the reptiles of the Pennsylvania Age of the upper Paleozoic, courtship ritual and sexual selection emerged. This was the beginning of a type of behavior that culminated in the dramas of Shakespeare.

Note that mate selection has developed in both the protostomes and deuterostomes. Since the last common ancestor of these two major groupings did not show sexual dimorphism and were often hermaphrodites, this is a case of convergent evolution. Given the emergence of sexuality in the earliest eukaryotes, there seems to be a tendency toward sexual dimorphism, which ultimately results in mate selection, once cognition emerges. There is something more general here than we have seen. Mate selection is clearly a determining feature in evolution.

The reptiles underwent an evolutionary radiation leading to dinosaurs, lizards and snakes, crocodiles, turtles, beaked reptiles, and the synapsida group leading to pelycosauria, therapsids, and stem mammals. The therapsids, known only through fossil remains, are regarded as reptiles that diverged in a mammal-like direction.

Summing up, beginning some 300 or more million years ago, the first reptiles appeared. Mammal-like reptiles began in the Permian Age, 280 million years ago. The therapsids were probably dominant from 280 to 225 million years ago when dominance shifted to the archosaurs including the giant dinosaurs. At 65 million years ago some massive climatological chance led to dinosaur extinction and the emerging mammals moved into all the vacant niches. The birds also emerged from the reptiles and radiated into the niches vacated by the extinction of the dinosaurs.

From the first amphibians to the first reptiles took about 80 million years. Along the evolutionary pathway from fish to amphibians to reptiles to mammals, there is a continuous growth and complexification of the brain. This occurs in a regular fashion, and one can map brain structures from taxon to taxon in a systematic way.

### Chapter 20—Readings

Cogger, Harold G., and Zwerful, Richard, 1998, *Encyclopedia of Reptiles and Amphibians,* Academic Press.

# 21

# Mammals

This chapter is concerned with the emergence of mammals from their early reptilian ancestors. The stem reptiles, descendants of amphibians, have been at the base of a radiation leading to turtles, snakes and lizards, crocodiles, and therapsids, which led to stem mammals, and the ruling reptiles, which led to birds. Since this radiation occurred early, we here concern ourselves only with emergence of mammals that include these features: internal temperature control, possession of mammary glands in females, and ability to give birth to live young, except in the case of monotremes, such as egg-laying mammals.

From fossil remains the dating of mammalian evolution (shown in Table 5 below) has been established.

The evolution from reptiles to mammals is a continuum, studied by analyzing jawbones and teeth because of the persistence of these structures in fossil remains. From the mammal-like reptiles, the cynodontia, the mammals emerged sometime during the Triassic Age. The transition from the first mammal-like reptiles to clearly recognized mammals took a very long time. One feature that is intensively investigated is the progressive enlargement of the jawbone (the dentary).

Unseen in the fossil record there must have been many changes in food gathering and efficiency of metabolism, mostly related to temperature homeostasis. Some students of dinosaur life believe that temperature homeostasis may have occurred independently in several groups of the ruling reptiles; it certainly occurred independently in the emergence of the birds. These kinds of evolutionary convergences are not uncommon. In any case, both mammals and birds exhibit temperature homeostasis.

TABLE 5: EVOLUTION OF TERRESTRIAL FORMS

| Millions of years ago | |
| --- | --- |
| 24-5 | Most mammalian families |
| 54-37 | Origin of most mammalian orders |
| 65 | Dinosaur extinction—mammalian radiation |
| 225-65 | Age of dinosaurs |
| 280 | Mammal-like reptiles |
| 320 | First reptiles |

There is an unresolved question whether mammals arose from the reptiles one or several times. In the jargon of the field of mammology, this is known as the monophyletic or polyphyletic origin of the mammals. In the first view, reptiles gave rise to early mammals, which then split into two groups, the monotremes and a larger group of marsupials and placentals. In the polyphyletic theory, there separately evolved from the reptiles a group leading to the monotremes and a separate group giving rise to the therians, both marsupial and placental.

During the great age of dinosaurs, the ruling reptiles were large and presumably diurnal, while the emerging mammals or mammal-like reptiles were small and nocturnal. With little niche overlap, both major taxa could coexist in the same ecosystems. With the overwhelming extinction, presumably due to a huge meteoritic impact, 65 million years ago, there was a great mammalian radiation to fill the niches left vacant by the disappearing dinosaurs.

By the time the mammals became the dominant animals, they differed from the reptiles in a number of ways:

1. For the same-size organism, the mammals have a large complex brain compared to reptiles.
2. The presence of mammary glands provides a food source for the newborns and forces a close social relation between the generations. A mother must feed her own recognizable offspring or group of offspring.
3. Mammals are endothermic and have a high metabolic rate compared to the ectothermic reptiles. Per unit mass, they metabolize at about 10 times the rate of reptiles.
4. Mammals have a muscular diaphragm; reptiles do not. This is clearly related to the need for more rapid respiration.

5. Mammals have a four-chambered heart, compared to the three-chambered heart in reptiles.
6. Mammals are covered with hair—reptiles, with scales.
7. Mammals are viviparous; reptiles are oviparous, except for those taxa that retain the eggs to hatch internally and then release live offspring.
8. Red blood cells are enucleate in mammals and nucleate in reptiles.
9. There are major differences in the jaw, dentition, inner ear, and growth of long bones.

The transition from reptiles to mammals was enormously complex. The therapsid reptiles that presumably gave rise to the early mammals were a diverse group, and the mammals display a wide variety of forms. It is as if aspects of mammalness kept appearing in parallel along various reptilian lines and finally emerged along three lines as organisms we clearly regard as mammals. These are the monotremes, the marsupials, and the placentals. This should not be oversimplified. We are presenting examples of emergence, but these must be eventually explained, or a general theory of emergence must be developed.

In the 65 million years of mammalian radiation, an enormous variety of types have evolved. These range from the tiny shrews to huge whales; from the water-dwelling cetaceans to the desert-dwellers; from the subzero polar bears to the camels living in over 100°F temperatures.

The mammals have been an enormously successful taxon and have given rise to some of the most physiologically and behaviorally interesting species living on the planet.

Two features of the emergence might be stressed. Temperature homeostasis with all its devices for heating and cooling allow mammals to inhabit and thrive in almost every ecosystem and climate. Feeding the young by the mother involves a longer period of intergenerational activity that emerges in transmission of learned behavior patterns. Many mammals are highly social animals.

This brief examination of the very extended emergence of mammals gives us a chance to repeat the point that emergence is a process, not a thing. Although fossils of mammalian-type reptiles appeared in the first 30 million years of reptile existence, it took another 100 million years to fully cross the line between reptile and mammal. Other lines of reptile evolution (and there were many of them) took place in parallel with mammalian evolution, but with one exception—they all led to reptiles or to extinction.

That exceptional case led to the emergence of birds, animals quite different from either mammals or reptiles. Some evolutionary lines give rise to animals radically different from the ancestral forms. Those are all emergences. There is, however, great continuity in biochemistry, cell structure, and physiology among all the emergent taxa. At the molecular level, the great unity within diversity persists.

There are many similarities between mammals and birds. This is perhaps to be expected since they are both advanced reptiles.

## Chapter 21—Readings

Davis, David E., and Golley, Frank B., 1965, *Mammology*, Van Nostrand Reinhold Co.

Feldhamer, G. A. Drickamer, L. C., Vessey, S. H., and Merritt, J. F., 1999, *Mammology*, McGraw, Hill.

Kemp, T. S., 1982, *Mammal-Like Reptiles and the Origin of Mammals*, Academic Press.

## 22

# The Niche

We have from time to time mentioned the concept of "niche." This ecological construct is a part of the Principle of Competitive Exclusion. We may understand the next group of emergences most easily in terms of the ideas of theoretical ecology. To establish some necessary background, we look back to the 1920s, a time when a prescient group of individuals was laying the foundations of ecological thought.

Alfred Lotka, the first of these investigators, was trained originally as a physicist and spent his career applying the mathematical methods of theoretical physics to a wide variety of other disciplines. In 1922 he went to Johns Hopkins University, where he spent two years writing *Elements of Mathematical Biology,* a book that has become a minor classic in theoretical biology. Substantial sections of the work were devoted to a mathematical analysis of the growth of populations under varying conditions. This analysis is an important basis of the competitive exclusion concept. (After completing his major scientific work, Lotka spent the remainder of his career developing actuarial concepts.)

At the same time that Lotka was carrying out his studies, an independent formulation of the mathematics of population growth was being postulated by the far-seeing Italian mathematician, Vito Volterra, of the faculty of the University of Rome. His extensive results on population lie at the roots of ecological theory in a manner analogous to Lotka's studies. They each modeled population growth by a family of differential equations. The works of the American and Italian theorists are so closely related that the fundamental mathematical relations have come to be known as the Volterra-Lotka or Lotka-Volterra equations. These equations led to a strong conclusion: when two species in the same habitat strongly compete with each other for some resource, one species wins out and the population of the other species goes to zero. Species and competition were rather formally described in the equations, species by assigning different independent variables and competition modeled by establishing the values of interaction parameters. A less abstract biological reference was necessary to give full meaning to the symbols and test the experimental validity of the equation.

The exact meaning of "species" of animals has been somewhat uncertain. Even Charles Darwin in his great *Origin of Species* wonders about the meaning of this term. For most animals we have come to consider a species as a group of organisms with a common genetic pool. Two groups of individuals will be considered members of separate species if they do not interbreed. For closely related species they might not be interfertile but do not mate because of behavior, habitat, or other biological constraints. The key to defining an animal species in a modern genetic context is in its reproductive interactions.

A fuller understanding of the meaning of competition came from the work of a young experimental biologist who was doing field work in the Arctic, observing its plants and animals at the same period that Lotka and Volterra were at their writing desks struggling with a set of difficult differential equations. In 1926, Charles Elton set down the thoughts that had emerged from his field studies and his deep but nonmathematical analysis of populations of plants and animals. At the early age of 26 he wrote *Animal Ecology*, a book that rapidly became a classic in a newly maturing science. Elton was responsible for formulating seminal ideas such as food chains, food cycles, niches, and the pyramid of numbers. We are here primarily concerned with the idea of a niche, which Elton introduced in an almost casual way:

> It is therefore convenient to have some term to describe the status
> of an animal in its community, to indicate what it is **doing** and not
> merely what it looks like, and the term used is "niche." Animals
> have all manner of external factors acting upon them—chemical,
> physical, and biotic—and the "niche" of an animal means its place
> in the biotic environment, **its relations to food and enemies.** The
> ecologists should cultivate the habit of looking at animals from
> this point of view as well as from the ordinary standpoints of appearance, names, affinities, and past history. When an ecologist says,
> "There goes a badger," he should include in his thoughts some

definite idea of the animal's place in the community to which it be-
longs, just as if he had said, "There goes the vicar."

Niche includes address and occupation. As the concept of a niche developed, it included the physical and biological description of an organism's
habitat, the role of the species in the biological community, and the resources that it utilizes.

With Elton's concept of a niche, competition could be better defined; it
was the result of two species trying to occupy the same niche. The strengths
of the competition depended on the similarity of the niches. The mathematical result of Lotka and Volterra could now be worded in the impos-
sibility of two species occupying the same niche in a steady-state
ecosystem.

That idea had also been seen from another point of view by Joseph
Grinnell, Director of the Museum of Vertebrate Zoology at the University
of California during the 1920s. A field biologist, Grinnell traveled widely
through the biologically diverse state of California collecting and studying
the fauna. From these field investigations he recognized that two species
cannot occupy the same niche. The principle of competitive exclusion,
which has many names, is sometimes called "Grinnell's axiom" in recognition of his early statement of the idea.

From field studies the principle of competitive exclusion moved into the
laboratory of the Russian biologist, G. F. Gause, at the University of Mos-
cow. He grew the unicellular animals paramecium in test tubes and regularly took them out and counted them. In the presence of abundant food,
he grew combined cultures of a large, slow-growing species *Paramecium
caudatum* and a smaller, faster-growing species *Paramecium aurelia*. In all
of his tubes both species grew and then the population of *Paramecium
caudatum* declined to zero while its close relative *Paramecium aurelia*
thrived. When he changed the conditions of the niche to make it more
favorable for *caudatum,* it thrived and *aurelia* perished.

Gause tried the experiment with various pairs of paramecium and also
extended his experiments with pairs of yeast, always with the same result
until he tried to grow *Paramecium aurelia* with *Paramecium bursaria*. Here
both survived at about half the population level attained when living alone.
It looked like a clear falsification of the principle until the experimenter
took a careful look and noticed *P. aurelia* living in the top half of the
growth tube and *P. bursaria* living in the bottom half. The tiny microcosm
of a single test tube had been divided into two niches by the behavior of
the survivors of the two original cultures.

Gause affirmed that in the laboratory the conclusion of the Lotka-Volterra equations were correct. As a tribute to Gause's work the principle of competitive exclusion is most widely known as Gause's principle. The studies outlined above appeared in 1934 in a book appropriately named *The Struggle for Existence*.

Following Gause's work, a number of investigators went back to the field to look for apparent exceptions to closely related species living in this same habitat. Such pairs are designated as sympatric species. An example is provided by the work of British ornithologist David Lack who studied two species of cormorants, birds that nested in the same cliffs and fed in the same area of ocean. Careful study showed that one species eats shrimps and nests on high cliffs and broad ledges, while the other eats sand eels and sprats and nests on low cliffs and narrow ledges. So, in spite of many obvious similarities the two birds do not compete: they occupy different niches. There are now hundreds of such cases, and competitive exclusion has taken its place as a major ecological principle. Where two species strongly compete, that is, try to occupy the same niche, then one will become extinct, or one will be driven out to try to find a niche in some more favorable habitat, or one or both will evolve to split the habitat up into two niches.

> In 1958, G. E. Hutchinson offered a more formal definition of the niche concept: The intensive definition required may be obtained by considering a hyperspace, every coordinate $(X_1, X_2, X_3 \ldots)$ of which corresponds to a relevant variable in the life of a species of organism. A hypervolume can therefore be constructed, every point of which corresponds to a set of values of the variables permitting the organism to exist. If no competitors are present, the hypervolume will constitute the **fundamental niche** of the species. If a number of species are living together but competing, each will have a **realized niche** usually corresponding to a smaller hypervolume than the fundamental niche, but by the principle of competitive exclusion, no point in one realized niche is also in another. This presentation allows for the fact that the direction of competition can change with changing environment conditions.

The niches may be in the canopy of tropical forests or in coral reefs or savannahs, but the notion of niche is essential to ecologically oriented evolution. Competitive exclusion is a major pruning rule in the emergence of biological taxa.

## Chapter 22—Readings

Gause, G. F., 1934, *The Struggle for Existence*, Williams and Wilkins.

Hutchinson, G. E., 1965, *The Ecological Theater and the Evolutionary Play*, Yale University Press.

Lotka, A., 1956, *Elements of Mathematical Biology*, Dover Publications.

# 23

# Arboreal Mammals

Several themes resonate through our chapters on biological emergence. These include coevolution, niche theory, and the growth of cephalization and mind in animal lineages. A fourth theme relates biological and geological events on a planet with shifting tectonic plates and the recycling of land and water. All of these events are curiously related.

The geological as well as biological conditions about 400 million years ago provided the conditions for the invasion of land by both plants and animals. The presence of insects and various plants was a prerequisite to the vertebrates coming ashore to find food and thus led to the amphibian radiation.

Following the initial development of terrestrial plants that were probably moss-like, three major groups of plants evolved. The earliest were probably the ferns, followed by the gymnosperms in the period from 363-286 million years ago. This coincided with the earliest land vertebrates. The gymnosperms are mainly large woody trees, of which the conifers are the dominant members. About 150 million years ago, the angiosperms or flowering plants arrived upon the scene. There are now some 200,000 species of angiosperms, and it is likely that the earliest forms were woody trees. Present species cover a size range from giant trees to tiny pond plants.

The presence of trees led to forests and a previously unrealized richness of niches. The many small and highly differentiated spaces in the plant world create distinguishable niches for small animals; hence their enormous variety, and the fact that insects have more species than any other taxon.

For the next emergence, we go back to a time at the end of the Mesozoic era and the beginning of the Cenozoic era, about 65 million years ago. The dinosaurs have gone extinct, and the remaining mammals, mostly small animals, are at the beginning of a grand evolutionary radiation. The birds have survived the extinction and, with mammals, are expanding into niches left vacant by the death of the ruling reptiles. The rise of the great angiosperm forest provided a whole series of novel niches. The emergence of arboreal mammals clearly depends on the emergence of forests. I repeat: all evolution is coevolution.

Working back to the roots of primate evolution, paleontological evidence seems to establish that primates evolved from insectivores. Since so much of early primate evolution depends on arboreal adaptation, it seems reasonable to investigate arboreal insectivores. The evolutionary development of grasping extremities including the tail seems like a favorable feature of mammalian invasion of the arboreal habitat, although bats and sloths have exploited other means. In any case, we move from claws and hooves to hands and feet. The tails also function in arboreal locomotion.

A number of characteristics of the arboreal mammals have gone on to develop further in the primates, as Matt Cartmill discusses in a chapter called "Arboreal Adaptations and the Origins of the Order Primates." (Tuttle 1972) The features of main interest are:

1. Grasping extremities
2. Olfactory regression
   (Lessening the importance of smell relative to vision)
3. Orbital convergence
   (Bringing of the eyes from the side of the head to the front)
4. Neurological integration of the visual fields

Cartmill points out that the arboreal environment is discontinuous, mobile, variable, and oriented at all angles to gravity.

Olfactory regressions mean a shift in brain and sensory function from smell to vision. The importance of vision in the trees is rather obvious, and the part of the brain allocated to vision as well as the size of the brain itself were involved in this evolutionary shift.

Orbital convergence. If one examines most mammals and the fossil skulls of mammals, the eyes are on the side of the head, providing for a 180° field of vision. In the arboreal mammals, the eyes move toward the front, and the snout recedes. This is a necessary condition to optimize stereoscopic vision. The ability to sense depth or distance is vital in navigating between branches. This ability is enhanced by the neurological integration of the visual fields.

The ability to live in trees introduces into the underlying mammalian genome a variety of features that later emerge as important in the hominids.

Another feature comes into play in thinking about arboreal mammals. This is the mental or noetic aspect of all animal life. Descartes in the late 1600s argued that animals had no souls and hence no minds. They were biological machines. The behavioral psychologists in the 1920s, 1930s, and 1940s had the same view. There is currently a reexamination that argues that mental activity is universally distributed through the animal kingdom and perhaps in other taxa down to the unicellular eukaryotes. Psychologist Donald R. Griffin has gathered a great deal of evidence in the book *Animal Minds* and argues for the universality of cognition. He discusses insects, birds, crabs, elephants, primates, and all manner of animals. All the studies point to cognitive abilities on the parts of animals. Something like mind seems to be present throughout the animal world. At the level of great apes, whales, and dolphins, he finds the evidence overwhelming. As we move from arboreal mammals to the primates and higher primates, we find more and more mental features emerging. The mental permeates the world of mammals.

I do not intend to present the emergences with a view to linearity, as if the unfolding of human society were a divine imperative that obscured all others. Each emergence is in a highly interconnected world of other interacting emergences. So, I am always at risk of being a bad biologist, a criticism also leveled at Teilhard de Chardin. Having stated that, I must confess that at a certain stage of the unfolding of life (I believe it is at the level of the arboreal mammals), I see the grand dawn of the emergence of reflective thought.

Reflective thought is not mind per se, for mind is almost a necessary part of animalness, as Griffin has so eloquently argued. Central here is the development of the aspect of mind that ultimately becomes the epistemic core of the hierarchical series of emergences. If emergences are a very large, highly interconnected network, we are investigating one pathway through the network. This privileged pathway leads to the possibility of the network's knowing itself. As George Wald once said, a physicist is the atoms' way of knowing the atom.

## Chapter 23—Readings

Cartmill, Matt, 1972, Arboreal Adaptations and the Origin of the Order Primates, in *The Functional and Evolutionary Biology of Primates,* Russell Tuttle, ed., Aldine/Atherton.

Griffin, Donald, 1992, *Animal Minds,* University of Chicago Press.

Hutchinson, G. Evelyn, 1965, *The Ecological Theater and the Evolutionary Play,* Yale University Press.

# 24

# Primates

Primates appeared on the scene about 65 to 70 million years ago, descendants of arboreal insectivores. They have been extraordinarily successful, giving rise to a large number of species, including some of the most interesting to study. The same basic body plan has been maintained from the 100-gram mouse lemur to the 200-kilogram male gorilla. They inhabit almost all of the tropical world and much of the temperate world.

The predecessors of the primates were probably similar to the modern-day tree shrews of the order Scandentia and the family Tupaiidea. They superficially resemble squirrels and are half a meter long from snout to tail. The limbs have five digits, with long claws and a large brain case. They may be regarded as evolutionarily between the shrews and the primates.

The evolutionary trends along the lines that define the primates are: refined hands and digits with nails instead of claws, elaboration of the cerebral cortex, and shortened muzzle. Sexual maturity occurs later and life spans increase.

Within a few million years after the arrival of the first primates there was a major split leading to the lemuroid stock and the tarsier/simian stock. Much of primate evolution has to do with response to changing habitats. The changes were due to plate tectonics and the breakup of the supercontinent into a group of smaller continents and islands. Again a biological emergence was driven by geological events.

The tarsiers and simians went their separate ways about 55 million years ago. The tectonic separation of South America from the rest of Pangea resulted in the Old World simian stock and the New World simian stock developing as separate taxa. The Old World stock is also called the Catarrhini, since the nostrils point down. It includes the Old World monkeys, apes, and humans. It seems clear that the monkeys preceded the apes and that branching occurred about 35 million years or less into the past.

Along with the primate body plans and prolonged adolescence, there emerged a social life that followed along into the next and successive emergences. The body plans are evolutionary consequences of adapting to the many niches in the tropical forest canopy. Primates are usually gregarious and live in groups ranging from a few animals to several hundred. The reasons for the competitive success of group living are:

1. Greater protection from predators. Groups may have specialization of labors with some individuals serving as guards.
2. Increased skill in securing and distributing food. This is particularly true in group hunting.
3. Access to mates. This shows considerable species variation.
4. Assistance in caring for offspring. Primate young are born at a quite immature stage. They require long periods of care, nursing, and training by the mother. The social system must permit this activity and free the mothers from other group activity.
5. Thermoregulation. This is seen in terrestrial primates such as baboons and macaques that huddle when sleeping.

The kinds of social groups most often seen in primates are:

1. Several adult females and males and their offspring.
2. A single adult male, several females, and their offspring.
3. A mated pair and their young.

The monkeys as forest-dwellers evolved a number of features of major fitness value that influenced the next emergence, the apes. We shall discuss some of these in the next chapter. Foremost among these primate properties was behavioral plasticity associated with brain development. This allowed them to function as generalists and to respond to change on a time scale fast compared to normal evolutionary genetic change. This behavioral plasticity is at the root of the ultimate human emergence as "rational animals." It allows acquired (learned) features to be transmitted, and thus introduces a Lamarckian transmission across the generations.

I can recall my first visit to a zoo, almost certainly the Bronx Zoo. For me the visit to the primate house was the high point. Watching the monkeys with their five-fingered hands and extraordinary manipulative ability gave me an eerie feeling. These were clearly our close relatives. Although I did not know much about evolution at the time, my zoo visit made me a committed Darwinian. Had I been aware of the term, I would have said I had had a conversion experience.

It is not only that we look like our relatives, but we use our visual and tactical systems in very similar ways. The emergence of primates must stand as a major step on the way to the noosphere. Life in the trees provided selection of many properties that ultimately emerged in higher primates. To take one example, the clinging of the young to their mothers was a selection for survival living in the branches. A result was a relationship between the generations that provided extended teaching and learning.

The emergence of primateness is not governed by a single rule, but by a complex series of selections associated with aboreal existence, making this emergence ecologically governed. Features of the brain then evolved that allow new kinds of primate behavior, cognitive as well as genetic.

### Chapter 24—Readings

Fleagle, John H., 1999, *Primate Adaptation and Evolution,* 2nd edition, Academic Press.

# 25

# The Great Apes

Between 20 and 30 million years ago, probably in Africa, a new group of primates evolved, branching off from the existing primates, the Old World monkeys. The second taxon, the great apes, emerged as a new type of animal in the world. That novelty has forever changed our planet. The most primitive of the great apes, the gibbons, branched from the main evolutionary line many million years after the first appearance of the apes. (Some biologists classify gibbons in a separate group, the lesser apes.) Some six million years after the gibbons, the orangutans, another group of tree-dwelling apes, appeared on the scene. Further evolution of the ape line starting about seven million years ago has resulted in present-day gorillas, bonobos, chimpanzees, and humans. The great apes are our branch on the tree of life.

How close are we to our primate relatives? With DNA technology, that question can now be addressed with numerical precision. The DNA in the genome of *Homo sapiens* (humans) and *Pan troglodytes* (chimpanzees) shows an overlap of 98.4 percent, or a difference of 1.6 percent. There is a similar overlap between humans and bonobos (*Pan paniscus*). The human/gorilla difference is 2.3 percent and the human/orangutan difference is 4 percent. Within the hominid groups to be discussed in the next chapter, it seems apparent that the differences are less than 1.6 percent, probably much less. On the basis of DNA comparisons, primatologist Jennifer Lindsey has concluded "that humans and great apes are more closely related than either is to monkeys." Humans are more closely related to chimpanzees and bonobos than gorillas.

The properties distinguishing great apes from monkeys are: taillessness, larger size, upright posture, broad chest, short snouts (noses), and larger brain-to-body size ratio. Something other than these physical measures has emerged along the great ape line. Cognitive and behavioral features appeared that make the superfamily Hominoidea unique in the world of living beings.

Great apes have a long period of maternal care. Chimpanzees begin to wean their young at three years of age, and orangutans at six. This means that great apes spend a long time being cared for and educated by their mothers. The genetic transfer of information is thus augmented by a transmission of learned information. A society of individuals living in close interaction is a major feature of most of the great apes. The group sizes may range from 2 to 20, depending on species and habitat.

The loss of the tail seems to be improved engineering for locomotion on the ground. Larger overall size permitted a greater range of mobility and increased protection in the world of tooth and claw. And so the other physical changes can be interpreted in a Darwinian fashion.

The ideas of emergence along the anthropoid line were stressed by Teilhard de Chardin in *The Phenomenon of Man:*

> So we have now before us, simultaneously with the true definition of the primate, the answer to the problem which led us to study the primates. "After the Mammals, at the end of the Tertiary era, where will life be able to carry on?"
>
> What makes the primates so interesting and important to biology is, in the first place, that they represent a phylum of *pure and direct cerebralisation.* . . . [E]volution went straight to work on the brain, neglecting everything else, which accordingly remained malleable. That is why they are at the head of the upward and onward march towards greater consciousness. . . .
>
> Hence this first conclusion that if the mammals form a dominant branch, *the* dominant branch of the tree of life, the primates (i.e., the cerebro-mammals) are its leading shoot, and the anthropoids are the bud in which this shoot ends up.
>
> Thenceforward, it may be added, it is easy to decide where to look in all the biosphere to see signs of what is to be expected. We already knew that everywhere the active phyletic lines grow warm with consciousness towards the summit. But in one well-marked region at the heart of the mammals, where the most powerful brains ever made by nature are to be found, they become red hot . . .

> We must not lose sight of that line crimsoned by the dawn. Af-
> ter thousands of years rising below the horizon, a flame bursts
> forth at a strictly localised point.
> Thought is born.

Returning to our geneology, about 20 million years ago or so, as we have noted, a new group of simians arose—the apes. The two present families of these primates are Hylobatidae and Hominidae. The first consists of gibbons and siamangs, and the second of orangutans, gorillas, chimpanzees, bonobos, and members of the genus *Homo,* whose emergence will be discussed in the next chapter.

The general view is that the apes evolved from Old World monkeys in areas where decreased rainfall converted forest into savannah. This placed a premium on walking and brachiating and decreased the value of the tail. There were fewer niches in the savannah, compared to the rain forest, and therefore competition may have been more severe, placing a premium on size and aggressiveness.

In any case, before the appearance of the *Australopithecus* (ape-man) lines, on the way to humans, five to ten million years ago, there were populations of tree-dwelling and savannah-dwelling apes with a high degree of manual dexterity, a substantial cranial capacity, and a social order characterized by teaching between mother and offspring and social learning. This is the beginning of something novel: cultural transmission of information. While the isolation of somatic cells and germ cells prevents the inheritance of acquired characters, the opposite is true of learned information that parents teach to offspring. In this domain one moves from a Darwinian to a Lamarckian transmission between the generations.

These features of passing on learned information did not begin with the higher primates, but certainly have developed most extensively in these taxa. The ability to transmit learned information clearly has survival value and will be selected for. We are moving toward a definition of fitness in which the pruning is a combination of biological and social. With the arrival of *Homo sapiens,* it becomes even more social.

Following the publication of Darwin's *Origin of Species* (1849) and *Descent of Man* (1871), great conflict broke out between the supporters of evolution and the traditional believers of biblical religion. The latter were especially horrified at the thought of human descent from apes. They had grown up in a long tradition of the special creation of man as the supreme act of an anthropocentric God. The 98.4 percent overlap in DNA sequence between humans and chimps leaves no doubt as to our common descent that is verified by every study on the anatomy, physiology, and behavior of these taxa. Special creation of man is simply an untenable hypothesis in any sense other than a unique series of emergences that will be discussed in the following chapters.

This is not to assert that there is not something very special about the appearance of thought. It is different, but it occurs within the evolutionary unfolding of the universe, not in opposition to it. *Homo sapiens* are a very special part of the natural world and are embedded in the noetic world. Nonetheless, the species has arrived by a series of emergences going back to the Big Bang. The value of thinking in terms of emergence is that it frees us from separate creation, but also notes the great chain of emergences that had to occur to arrive at entities that can look back and try to understand the emergences.

## Chapter 25—Readings

De Waal, Frans, and Lanting, F., 1997, *Bonobo, the Forgotten Apes,* University of Chicago Press.
Diamond, Jared, 1992, *The Third Chimpanzee,* HarperCollins.
Lindsey, Jennifer, 1999, *The Great Apes,* Metro Books.

# 26

# Hominization and Competitive Exclusion in Hominids

Among the great apes, a subgroup diverged about five or six million years ago that we designate as the hominids. Our knowledge of this group comes from the lone surviving species *Homo sapiens* and an array of fossilized bones and some artifacts such as chipped stones, burial items, and other remains that have survived the decay of time.

In this chapter we will briefly review the five million years from the first man-apes to *Homo sapiens*. The entire emergence is designated hominization. We shall then go back from the big picture to look at several emergences along the way. As in other cases, there is not a unique set of specific emergences, but a continuum, which brought us from the savannah- and jungle-dwelling great apes to modern humans.

I think we have already made two things clear: first, we are animals whose physiology, anatomy, and biochemistry places us clearly among the mammals, operating in every way as inheritors of four billion years of molecular biology utilizing the same rules as are common to all living forms. Second, several differences have developed along the anthropoid line, culminating in a species quite different from any other, a species with great capacity to change the environment, and a species that attempts to understand itself and its surroundings. We are a species that constructs its own past.

The hominid line is characterized by upright bipedal locomotion and an enlarged cerebral cortex. The oldest known potential member of this branch of the evolutionary tree is *Ardipithecus ramidus,* represented by some fragmentary fossils from the 4.4-million-year-old site of Aramis in Ethiopia. (*Scientific American* 282–7, 56-62, 2000). The related species of

*Australopithecus* and *Paranthropus* are found from 4.2 to 2.5 million years ago.

The early species were all hunter-gatherers, and there is not much to indicate any material culture developing in this period, although the evidence is minimal.

About 1.8 million years ago, East Africa seems to have been the home of at least four species of hominids: *Paranthropus boisei, Homo rudolfensus, Homo habilis,* and *Homo ergaster.* These species seem to have replaced the Australopithecenes who earlier inhabited the African continent. The 98.4 percent overlap between *Homo sapiens'* DNA and chimpanzee and bonobo DNAs suggests an even closer resemblance among the hominids. With *Homo habilis* from Koobi Fora and other sites in East Africa, there is evidence of toolmaking, particularly involving the chipping of rock. Indeed, a suggested factory site has been found at Koobi Fora. With the appearance of the genus *Homo,* there appears an extensive migration of hominids reaching to China and Java and various locales on the European continent.

The paleontologist Ian Tattersall suggests that the past million years led to a radiation in locale and species involving *H. erectus, H. antecessor, H. neanderthalensis, H. heidelbergensis,* and *H. sapiens* about 25,000 years ago. The sole surviving species is *Homo sapiens.*

The evolutionary radiation of hominids can probably be explained by isolation of small groups, local genetic changes, and selection for fitness in various habitats. The early hominids were quite free ranging and could respond to competition by moving on. Gause's principle that no two species can occupy the same niche is as valid for hominids as for any other species. The principle results is one species killing off the other or by one species leaving the niche.

To spread out from East Africa to China in 10,000 years requires an average rate of 1.5 kilometers per year, or 15 kilometers per generation, a modest rate indeed. The net result was a radiation in space and species.

Eventually population increase expanded the physical size of niches, and migration was replaced by some sort of warfare as the dominant struggle for niches. To this day, this partially explains one race, religion, or ethnic group's killing off another. Let us then look at Gause's principle of competitive exclusion in more detail.

The Principle of Competitive Exclusion, deep within the structure of modern ecological theory, is still applicable for *Homo sapiens.* It has been possible to examine applications of this principle in human social behavior. Competitive Exclusion illuminates discussion of political problems for societies in which two distinguishable groups try to coexist. It may also illuminate the rise of hominization. Understanding competitive exclusion may be the key to realistic attempts to deal with conflicts in Northern Ireland, Sri Lanka, Israel, Fiji, and certain other locales such as Belgium where serious intergroup conflicts usually lie just below the surface and occasionally break through.

Having previously pointed out the generality of competitive exclusion, we next move to a case in modern human societies that is closely analogous to the condition of this ecological theory. In a number of contemporary societies two or more groups live in the same country and exist in almost complete reproductive isolation because of religion, race, language, ideology, or other nonbiological barriers. As a result an analog to sympatric species is artificially produced. We designate these noninterbreeding groups as pseudospecies. Indeed, if such isolation were complete enough and lasted long enough one might anticipate over a very long time that speciation might occur. Human societies have not been stable for sufficiently long periods to test this hypothesis, but clearly certain genes such as Tay Sachs and sickle cell anemia are associated with certain identifiable populations in the world. If the genetic divergence became sufficient, speciation could in principle result in the absence of interbreeding. The strength of the human sex drive seems to prevent this and maintain the flow of genes.

It is abundantly clear at this point that all humans are members of the same species *Homo sapiens*. Nevertheless, interfertile but relatively isolated groups may have experienced genetic selection for particular characteristics. Compare for example the Watusi of Burundi with the not-too-distant Mbuti pygmies of Zaire. These groups vary in average height by at least 12 inches. But in spite of the close biological affinity of all humans—according to mitochondrial DNA studies we all have a common ancestor within the last 200,000 years—humans constantly erect cultural barriers to interbreeding, where no biological barriers exist. These can result in measurably different physical characteristics, as in the case of the Watusi and Mbuti mentioned above.

Where two culturally non-interbreeding groups or pseudospecies live in the same area and compete for a common resource, the conditions of the exclusion principle exist and strong competition, according to the Lotka-Volterra analysis, should lead to the extinction of one group. Now, humans, having reflective thought and the power of choice, are not bound to living out a set of mathematical relations, and we shall examine some possible social and political responses to a socially imposed problem.

One of the solutions to competitive exclusion is the formulation of non-interbreeding groups or castes, which are tied to occupations or roles in the society. This association between caste and means of making a living creates separate niches by removing competition for resources. In classical India there were 3,000 castes grouped into four main groups, and in addition there were the untouchables. Groups were associated with functional niches to avoid competition. Some residuum of this still survives in this part of the world.

In medieval Japanese society the Eta were a caste of butchers, tanners, and leather workers chosen from the general population by occupation. Obliged to marry only within their own group, the Eta became a pseudospecies. There is still a general unwillingness among the Japanese to intermarry with the Eta or indeed to discuss the problem. The caste status has persisted over a hundred years after all legal restrictions have been removed. This is an example of the remarkable persistence of some social barriers.

One response to competition is for one group to emigrate either voluntarily or by force. A modern example of this was the expulsion of 40,000 Asians (largely Indians) from Uganda in 1972. Independence in 1962 created competition between the Asian merchant class and other Ugandans. The expulsion was a particularly precipitous response brought about by the violent rule of Idi Amin. Tribal rivalry in East Africa is a long-standing example of niche competition.

Another response is to have sufficient interbreeding to remove the existence of pseudospecies behavior. In Hawaii, this solution has come about for a variety of reasons and one assumes that in a few more generations the original immigrant groups will be thoroughly mixed. However, there are still occupations (resource niches) that tend to be restricted to groups of national origin. James Michener in his book *Hawaii* has coined the phrase "golden man" to refer to the genetic mixture of Polynesians, Caucasians, and Asians that is emerging among the Hawaiian population. As a famous sociologist once noted, "It's not that the Hawaiians don't have prejudices, it's that they don't take them to bed with them."

On the American mainland the institution of slavery created a temporary caste system, which was suddenly completely changed by emancipation. This left two relatively noninterbreeding groups competing for jobs and other resources. While this has been followed by over a century of violence and attempts to maintain caste, one senses in the current milieu that the melting-pot philosophy will lead to sufficient interbreeding eventually to merge the two groups. This is, however, by no means certain.

Early in the seventeenth century Northern Ireland was inhabited by an indigenous-Irish, Gaelic-speaking, Roman Catholic population. Following the Tudor invasions there was extensive settlement by English and Scottish Protestants who became the majority and the economic upper class. Following the independence of the Irish Free State in the South, the non-intermarrying Catholics and Protestants found themselves in economic and political competition. Northern Ireland has become a classical case of competitive exclusion influenced by British military control to try to police the overt violence.

The situation in Sri Lanka is well illustrated by the following description from *Encyclopedia Britannica:*

> The population, while consisting of nationals of Sri Lanka, is divided into several groups. The two principal linguistic groups are the Sinhalese, who generally are Buddhists and number over 9,000,000 and the Tamils, who generally are Hindus, numbering more than 2,500,000. Muslims (misnamed Moors), who number over 850,000, also speak Tamil. Relations between Sinhalese and Tamils have deteriorated in recent years, more as a result of current economic maladies than for historic reasons. Linguistic and religious groups—as well as the caste groups into which both Sinhalese and Tamils are divided—vie with one another to secure economic and political advantage, using pseudo-historic race myths and concepts in their struggles.

Several thousand people have been killed in recent years in conflicts between Sinhalese and Tamils.

Similar situations exist between the Jews and Arabs in Israel and the West Bank, among the Fijians and Indians in Fiji, and the Flemish and French in Belgium. In all cases the two populations are maintained by barriers to intermarriage and interbreeding, and in all cases the negative effects of competition show up in stresses that threaten the very fabric of the society. By seeing these and a number of other cases within the context of Gause's principle one can better analyze the political options available to stabilize the societies. The principle also helps one identify proposed solutions that will not work because they leave two pseudospecies in strong competition.

Partition should work within the realm of competitive exclusion because it provides acceptable niches and therefore minimizes competition. The major partition of the Indian subcontinent into India, Pakistan, and Bengal eliminated much Hindu-Moslem competition, but did not solve the problems caused by certain separatist groups remaining in each country, as demonstrated in Kashmir and the Sikh separatist movement.

Of course the most humane solution is interbreeding. It removes the category of pseudospecies, relates individuals, and blurs the social barriers that separate individuals. The French and English populations of Canada represent a case where partition has been proposed, but a sufficient degree of interbreeding may exist so as to resolve the problem in a couple of generations or more. The large number of bilingual individuals certainly eases the situation. The outcome is still uncertain.

In all cases the most civilized solution would seem to be learning to live together in peace, though different. If competition for resources (jobs, economic advantage, land, etc.) continues to exist, then competitive exclusion suggests that this benign solution will *not* work. I suspect that truly living together in peace implies interbreeding, and if this degree of peaceful coexistence is withheld for whatever reasons, then a certain degree of hostility remains, to become overt when the competition gets tough. Not consenting to interbreed is in a sense the ultimate rejection: it sets people apart on the path to speciation. Ideologies and restrictive social practices, insofar as they cause pseudo speciation, are on the road to hostility. Forty or so years after the principle was shouted in the streets, "Make love, not war" seems like a solution that truly avoids competitive exclusion.

Competition for land and other resources nearly led to genocide in the case of native Americans on the mainland; it was even more genocidal in the case of the Caribs of the West Indies and led to compete genocide in the case of the Tasmanian aborigines. Apartheid in South Africa was an attempt to establish caste niches with vast inequities in the distribution of resources. It placed repressive rules on the competition for resources. By competitive exclusion, apartheid without partition had to have failed. Either whites and blacks will interbreed, or one group will in time go to extinction. Because of worldwide electronic communication, the caste solution was no longer socially acceptable, even though vestiges of it remain in many, indeed almost all, societies. International bodies such as the United Nations now reject making a caste system official government policy. The world response to South Africa's system exemplified the feelings about caste. However, even since apartheid has been abandoned, South Africa is still subjected to the consequences of competitive exclusion if barriers to interbreeding remain. An example is Zimbabwe, where attempts at biracial society are being thwarted with an extensive emigration of whites. How much interbreeding is required to break down pseudospecies has not been determined. It is subject in part to mathematical study and that analysis should be pursued.

It might be argued that regarding human beings as parameters in population equations ignores our ability to respond in more complex or humane ways. If we can create pseudospecies by a nonbiological or cultural device, can we not respond to competition in more imaginative ways? In principle, I suppose we can, but it is difficult to think of any device to reduce competition other than some form of caste.

Thus, I think we are left with the necessity of taking competitive exclusion very seriously as a sociological principle. Within this context political theorists and politicians must seek solutions compatible with this law of species or pseudospecies interaction. We cannot formulate a political or religious practice that violates our biological nature.

What then distinguished race from species in a developing radiation of hominids? In general, species membership is defined by free flow of genes. Interbreeding depends on anatomical, physiological, and behavioral factors, all of which are seen in the broad array of mating habits and rituals. In hominids it would seem that cognitive factors entered the distinction between self and not self.

In a battle between hominid species, the victors kill the vanquished. In battles between hominid races, the victors breed with the vanquished. The difference is a flow of genes between the groups.

These considerations of competitive exclusion give us some insight into the five-million-year saga from the first apes to walk erect to the emergence of modern *Homo sapiens*. The problem is complicated by the paucity of fossil evidence, the incompleteness of various skeletal remains, and the distribution of materials over three continents. Following Darwin's views, there was a tendency to seek a linear array from ape to ape-man to human. As the evidence begins to pile up, it seems that the radiation generated perhaps a dozen species. Some were wiped out in competition, others may have interbred before complete speciation, and others thrived and branched off new forms. The ability of small groups to travel to new habitats presented the opportunity for founder effects.

Certain features such as brain size and overall height seem to have systematically increased. The difference in size and physical features of various groups of present-day humans give us a clue to what may have happened. In any case, the emergence of a single-species *Homo sapiens* seems to have characterized the last 200,000 years. Competitive exclusion seems like an adequate pruning protocol for this emergence.


## Chapter 26—Readings

Gause, G. F., 1934, *The Struggle for Existence*. Williams and Wilkins.
Tattersall, Ian, 1998, *Becoming Human,* Harcourt Brace & Co.
Tattersall, Ian, 2000, *Scientific American* 282, 56–62.

# 27

# Toolmaking

Toolmaking is not a property unique to *Homo sapiens;* nevertheless, the emergence of certain sophisticated types of fabrication of devices appears as one of the emergences along the hominid pathway.

Chimpanzees carry large stones from several meters away and crack open nuts, clearly demonstrating tool usage in the anthropoid line. They also break twigs off a branch, strip the leaves, poke it in a termite nest, and bring it out covered with termites, a favorite food.

Herons drop small pieces of twigs to attract minnows, which they then swoop down to eat. Beavers are famous dam builders who also plug water leaks and make holes to lower the water level. Nest building by insects often involves considerable sophistication in gathering and placing materials. Tool usage is neither particularly common nor particularly rare in the animals.

Among hominids there is an association between tool type and evolving species. Tool usage is an emerging property of humans that began about 2 million years ago and has continued up to the present. The earliest surviving tools were chipped stone projectiles recovered at Olduvai in Tanzania, perhaps from as long as 2.5 million years ago. The use of chipped stone tools and the emergence of the genus *Homo* may be contemporaneous, with the suggestion that *Homo habilis* is that species.

After chipped stone, the next major technological advance is the hand ax. About 1.5 million years ago, *Homo ergaster* probably invented this device in East Africa. The advance in the manufacture of tools by stone chipping probably reached a high point in *Homo neanderthalensis* shortly before that group disappeared.

The ability to make tools paralleled advances in cranial capacities, although there is not a necessary relationship. Some average estimated cranial sizes are shown in Table 6 below. Although there is a general growth in size, the cranial capacity of *Homo sapiens* is not as large as that of the Neanderthals, although multiple factors are clearly involved.

Ian Tattersall has formulated an interesting theory about the relationship between *Homo sapiens* and *Homo neanderthalensis*. About 100,000 years ago, both species inhabited what is now Israel and appeared to co-exist with a similar level of material culture. Europe was inhabited only by Neanderthals with a level of material culture similar to both groups in Israel.

About 40,000 B.C. *Homo sapiens* invaded Europe. In 10,000 years, *Homo neanderthalensis* disappeared. The invaders were clearly different from the *Homo sapiens* that lived in Israel. Their material culture had enormously improved. They had improved stone working and made tools from bone and antler. They were musical and artistic, and their elaborate burials spoke to belief in an afterlife. In short, self-awareness and a high degree of social organization now characterized *Homo sapiens*. Modern man had emerged both biologically and socially in the 60,000-year period. Tattersall suggests this is the "emergence of modern cognition which it is reasonable to assume is the advent of symbolic thought."

Along with toolmaking, the emergence of language must have occurred in *Homo sapiens* between 100,000 and 40,000 years ago. Along with the slow emergence of toolmaking, which goes back into our animal past, the more dramatic rise of language and symbolic thought probably began about 60,000 years ago—only going back about 2,000 generations. Modern *Homo sapiens* are truly a recent arrival on our planet. The authors of the Old Testament were vastly off in dating the origin of the universe and the solar system at 6,000 years ago, but in dating the "creation" of man, they were only off a factor of ten or less, assuming that man as we know him is coincident with linguistic man.

The growth of toolmaking has been a constant feature from *Homo habilis* onward. The past 10,000 years has been the great age of tools. We now have in a sense produced the ultimate tool, the robot, and this tool has been fabricated with informational capacity. Some raise the question of whether the tool is now poised to take over from the toolmaker. This too is a potential type of emergence.

With the appearance of *Homo sapiens* various emergences are no longer independent, since social activities are complex and interdependent. Thus, toolmaking, language, agriculture, and war all relate to each other. We

TABLE 6: PRIMATE BRAINS

| Animal | Cranial Capacity |
| --- | --- |
| Orangutan | 400 cc. |
| Chimpanzee | 400 |
| Gorilla | 450 |
| *Australopethicus* | 450 |
| *Paranthropus* | 475 |
| Early hominids | 650 |
| *Homo erectus* | 975 |
| Archaic *Homo sapiens* | 1300 |
| Neanderthal | 1400 |
| Modern humans | 1200 |

treat them separately for convenience to assess the major features in a culture's coming to be and defining its characteristics.

Toolmaking emerged at least two million years ago in *Homo habilis*. A number of stone tools have been recovered from a wide variety of locations, and they constitute much of our record of hominid culture. These artifacts vary around the world, but for the Near East and Europe, 10,000 years ago is the end of the Old Stone Age (Paleolithic) and beginning of the New Stone Age (Neolithic). Starting in the Neolithic, toolmaking becomes technology and will be discussed in a subsequent chapter.

Hominid evolution can be mapped onto with the emergence of toolmaking. They may not be independent emergences. For as we get to the genus *Homo,* the rules change and biological evolution and cultural evolution become interdependent.

It is a long way from *Homo habilis* to Immanuel Kant, but something occurred in toolmaking that is a large part of how we have come to view the world in our philosophy of science. In the Kantian view, we start out with the mental: the *a priori* (basic principles) and the *a posteriori* (observations). We then form constructs that are the ultimate elements of science. The first construct is reification: we postulate that the stone is real and will be there when we shut our eyes and turn our head. Our next construct is that the stone has an inside that will appear as we chip away. The utility of chipping depends on a kind of causality, a belief that a certain kind of blow (cause) will result in a certain kind of chip (effect).

To be a toolmaker is the beginning of being a scientist. It is this epistemologically rooted thought process that ultimately makes up who we are and distinguishes us from other animals. It is not just that man is a thinking animal, but man is an animal who forms constructs to explain the observed world. It is not clear where this began, but it is a mental feature that became central in toolmaking.

## Chapter 27—Readings

McClellan, James E. III, and Dorn, Harold, 1999, *Science and Technology in World History*, The Johns Hopkins University Press.
Tudge, Colin, 1996, *The Time Before History*, Scribners.

# 28

# Language

There is little doubt that the emergence of language was a crucial step in hominization. There is considerable debate as to whether language was acquired over a long period of two million years, beginning with *Homo habilis,* or suddenly emerged in modern *Homo sapiens* some 60,000 years ago. For our form of language, the following features must have developed: (1) the ability to hear and process sounds, (2) the ability to produce a sequence of distinguishable phonemes, (3) the ability to assign meaning to the processed sound signals, and (4) a grammar to lead to a sophisticated language. We will discuss some of these features, with the understanding that much is missing in our understanding of language. Before writing existed, there is little in the way of paleontological records related to language. Speech, hearing, auditory processing, semantics, and grammar are all associated with soft tissues, which leave no fossil remains. In addition, language is a social, as distinguished from an individual organismic, activity.

Auditory processing ability clearly precedes language. A number of domesticated animals are able to distinguish phonemes. Dogs, cats, parrots, and apes can all respond to complex verbal instructions. Any habitat has a rich variety of sounds that are informative about the habitat, weather, and the presence of other animals. Being able to hear and distinguish all of these sounds is certainly a feature of fitness. The evolutionary development of hearing and auditory processing equipment by a large number of species confirms the generality of response to some signals.

A great number of animals can produce distinguishable sounds, but language to have any information density must utilize a substantial number of phonemes. One can see this using elementary information theory. If there are N possible phonemes, then, using elementary information theory, a message x phonemes long has an information content, I, of:

$$I = x\ln_2 N$$

The larger the value of N, the richer a language can be in terms of bits per phoneme, or the speed of transmitting information.

Of course, at issue is not only the number of phonemes an organism can make, but the number that it can make, hear, and process. We know that there are learning-disabled individuals with an auditory processing deficit who cannot distinguish all the phonemes that are used in normal human speech. If such an individual can only process M phonemes, the information per phoneme received is

$$I_r = \ln_2 M$$

And the information lost per phoneme is

$$I_L = I\text{-}I_r = \ln_2 (N/M).$$

This number measures problems of hearing and auditory processing.

The work of Savage-Rumbaugh et al. (*Language Comprehension in Ape and Child*) suggests that the auditory processing apparatus for hearing and processing language has been in place for at least six million years and predates hominid evolution. That is, both the hearing apparatus and the neural apparatus go back a long way. A series of other animals that respond to speech suggest this is much older. The best known example of a member of a remote taxon who responds to speech is Alex the parrot, studied in great detail by Irene Pepperberg in *The Alex Studies*.

The ability to produce a series of phonemes is found in the talking birds, but does not seem to occur in the nonhuman great apes or the other primates. Note that, from the information content per phoneme ($\log_2 N$), the more phonemes that can be produced, the more information-dense the language. Somewhere along the human evolutionary pathway, apparatus evolved that permitted, under appropriate neural control, the production of a large number of phonemes. The $\log_2 N$ value indicates that the number of phonemes is important in an information-dense oral language.

The next skill necessary for language is the ability to assign meaning to an array of phonemes. This is the subject of semantics and is still not well understood, for it is related to human thought and the problem of how a series of sensory events is transformed into thoughts. It is fair to say that changes in the brain must be associated with the ability to convert sounds to meaning, but this is poorly understood.

The next aspect of language is grammar, a rule set that governs the use of language and the generation of meaning from the message elements. There is an ongoing argument about whether grammar is taught or is wired in the neural elements of the brain.

Once human language emerged, there must have been an almost explosive radiation into a rich variety of phonemes and grammars. Any given language uses only a restricted set of phonemes. Grammars are quite variable, although there are certain uniformities. Languages have become enormously diverse, yet they are crucial for further emergence. Language is the medium of science, philosophy, religion, literature, and much of human social interaction. Note that human languages may use as few as 10 or as many as 60 phonemes.

Terrence Deacon speaks of the coevolution of language and the brain, and this must be true, in a certain sense. Language also relates to cultural evolution, a particular feature of *Homo sapiens* by which Darwinian evolution is replaced by changes in social patterns. Since culture lacks the distinction between germ plasm (sperm and egg) and somato-plasm (body tissue), which characterizes biology, acquired patterns of behavior can be transmitted, and human life moves into a Lamarckian domain.

Language is an enigmatic emergence. It is genuinely novel and is hard to relate to phenomena that preceded it, both from an anatomical and a cognitive point of view. It is clear that human thought and language are very closely related, but it is difficult to formulate the relationship. I recall struggling with a book called, *Why Does Language Matter to Philosophy?* by Ian Hacking. I read it several times, but the material remained enigmatic. Thought is so totally intertwined with language that it is difficult to separate them.

Language is part of what makes *Homo sapiens* the creative being that he is. We cannot date the emergence or even know the species that first utilized language. Language reminds us of what a social construct cognition is. The anatomical and physiological production and sensing of phonemes requires a minimum of two individuals. Our entire view of the world is embedded in language. Language presumably arose at a time between *Homo habilis* and Cro-Magnon *Homo sapiens*.

Because the history of language is so obscure, attempts have been made to study the emergence by viewing the process of how children acquire language. Studies of this problem are associated with Jean Piaget, Noam

Chomsky, and others. For the next few decades I expect that linguistics, phonetics, semantics, and their relation to neurobiology will be central areas in the study of cognition.

## Chapter 28—Readings

Deacon, Terrence W., 1997, *The Symbolic Species,* W.W. Norton & Co.

Hacking, Ian, 1975, *Why Does Language Matter to Philosophy?,* Cambridge University Press.

Pepperberg, I. M., 1999, *The Alex Studies; Cognitive and Communicative Abilities of Grey Parrots,* Harvard University Press.

Savage-Rumbaugh, E. Sue, 1993, *Language Comprehension in Ape and Child,* University of Chicago Press.

# 29

# Agriculture

At Jericho in the Jordan Valley is a mound, called a tel, built up over the Old City and comprising at least 25 layers of civilization. Near the bottom are remains of a farming village 10,000 years old. Jericho has the oldest remains of an agricultural society emerging from hunter-gatherers. The transition from the earlier to the later form of civilization could only have occupied a few hundred years, yet forever changed how humans live. It was truly a watershed event, yet not a unique one.

The past 50 years have seen intense study of the remains of the last 10,000 years of human habitation. Vastly improved carbon-14 dating methodology has lent a great precision to the study of remains of inhabited sites. Paleontology and paleobotany have provided abundant traces of early farming, the species employed, and other information about the plants and animals that were first domesticated. DNA sequence studies permit biologists to establish relations between domesticated species and their wild antecedents.

Ten thousand years is a short time span in the age of the planet. It is also just about three to four hundred generations in human dimensions, and it goes back to the beginning of history. Place as well as time is important. Agriculture began in the Fertile Crescent, the strip of land from Egypt to Turkey. Agriculture led to urbanization, which led to dynasties. The Fertile Crescent is also the home of Western religion.

When the authors of the Bible tried to chronicle the entire history of the world, they went back into the epoch of the emergence of agriculture as the "creation" they could relate to. The hunter-gatherers Adam and Eve gave rise to the herders and farmers Cain and Abel. The idyllic world of the hunter-gatherer changed to a world of working "by the sweat of one's brow." The biblical dating of creation is close to the emergence of agriculture.

There appear to be at least seven independent emergences of agriculture around the world as shown below in Table 7.

There were and perhaps are isolated peoples of the world who never developed agriculture on their own; yet, there is no reason to believe that they would not, in time, have discovered the cultivation of crops and domestication of animals.

There are also pre-agricultural practices. For example, Australian aborigines pull yams for food and plant the shoots to grow more yams for the next time they pass by.

*Homo sapiens* as a species is about 200,000 years old and has spread over much of the habitable world. Many groups have been isolated from others, as seen in the evolution of races. Yet within a 6,000-year window, agriculture seems to have been discovered (implemented) by seven very diverse and widely separated groups. To those who would attribute this to seven frozen accidents, I would argue that agriculture seems like an emergent social activity of *Homo sapiens* living in temperate and subtropical areas of the Earth in ecosystems with grains and herbivorous animals. Seven successes suggest that if the tape were played over again, the result would be agriculture. It is an emergent activity of societies of humans, but an activity of great consequences.

While the hunter-gatherer groups had 30 to 40 individuals, the early agricultural communities seem to have around 300 individuals. Settled habitations with year-round houses replaced mobile tents and other portable housing. Division of labor followed, and technology emerged.

In the Fertile Crescent considered as a whole, the transition from a hunter-gatherer economy to a farming economy occupied about 2,000 years. By 8,000 years ago, cattle, pigs, sheep, and goats had been domesticated, and barley, wheat, and lentils were regularly cultivated. By the end of the 2,000-year period, cities, city-states, and nations replaced the temporary dwellings of hunter-gatherers.

We do not know and probably shall never know whether population pressure led to agriculture as a solution to the need for food, or whether the invention of farming provided a setting for population growth.

Domestication provides for rapid evolution of the species employed. A selection for fitness was imposed by the domesticators. Thus, large seeds, fleshy fruits, and other desirable properties were selected in choosing for planting. In time this had its genetic consequences, and new strains

TABLE 7: EMERGENCES OF AGRICULTURE

| Area | Number of years ago |
|------|---------------------|
| Fertile Crescent | 10,000 |
| South China | 8,500 |
| North China | 7,800 |
| Central Mexico | 4,700 |
| Eastern United States | 4,600 |
| South Central Andes | 4,500 |
| Sub-Saharan Africa | 4,000 |

emerged. Similarly, in animal populations selection for docility or size or some other feature would alter a species. In *The Origin of Species,* Charles Darwin gave many examples of changes from selective breeding. He thus demonstrated the mutability of species that became part of his argument for the origin of new species. The rapidity with which breeds of dogs are developed is an example of the rapid changes possible for controlled varieties. In the case of corn (*Zea mays*) the alteration was so profound that the domesticated plant cannot reproduce in the wild because the seeds are not released.

The coevolution of domesticated plants follows a different set of rules from the evolution of plants in nature. In a few thousand years, large areas of grasslands and forests were converted to farmlands. Natural ecosystems were replaced by fields of single-species monocultures. Water runoff was altered. The lithosphere and hydrosphere were changed in major ways by agriculture. This emergence is the beginning of the conversion of the world of nature into a world totally dominated by human landscape.

The domestication of animals seems to have begun at the same time as the farming of plant crops. Animal domestication might have been preceded by hunter-gatherers who captured orphaned young animals and raised them in captivity. This would have provided information necessary for large-scale animal domestication. In a very real sense, agriculture is a knowledge-driven activity. Knowledge must be communicated by language and must eventually be stored by writing.

Although the beginnings of agriculture must have occurred in specific communities, in regions such as the Fertile Crescent, there was a rapid spread and synergistic interaction of farming and domesticated species. Agriculture by 8,000 years ago was an economic activity of the entire Crescent, the root of the culture of that entire part of the world. Parallel developments were underway in the Orient.

Agriculture is then an emergent social activity of *Homo sapiens* that leads to control of the physical and biotic habitat. The diversity of its occurrence in relatively noncommunicating parts of the planet suggests that it is a rule-driven emergence, selected for by increasing the average fitness of individuals in such societies. It is the beginning of a series of emergences that result in increasing human control over the environment.

All emergences raise the question of whether they are unique or multiple. Agriculture seems like a clear example of a multiple emergence. It encourages us to raise the questions of uniqueness about other emergences.

Why did it require 200,000 years for early man to develop agriculture? It may have been that for omnivores such as humans, the bounty of nature provided sufficient and reliable sources of food. It may also have been that an occupation as intellectually sophisticated as farming required a long cultural incubation, developing a language of abstract concepts such as germination of stored seeds and implementing the concepts in social activities.

The beginnings of agriculture also coincide with the boundary between the Pleistocene (Ice Age) and the Holocene (Recent Age). Changes in climate led to changes in vegetation and animal species, which may have been more suitable to agriculture.

The rate of change has sped up from the earliest hominids onto the cultural domain. Throughout the Holocene, the changes have been anthropogenic, caused by *Homo sapiens*. Something very major emerged about 10,000 years ago.

## Chapter 29—Readings

Smith, Bruce D., 1995, *The Emergence of Agriculture,* Scientific American Library.

# 30

# Technology and Urbanization

Technology did not suddenly emerge. In hominid society it goes back at least to the toolmaking *Homo habilis* living in East Africa and chipping away at stone rocks. The period from two million to 10,000 years ago is designated as the Paleolithic or Old Stone Age. Technological advance came slowly to small groups of hunter-gatherers.

Ten thousand years ago, coincident with the rise of agriculture, a new technology emerged that involved grinding and shaping stone as well as chipping. Tools appeared that used bone, wood, and hide as well as shaped stone. The emergent age is designated the Neolithic or New Stone Age. Thus the time 10,000 years ago (give or take a few thousand in different parts of the world) is a triple transition: from the Pleistocene to the Holocene, from hunter-gatherers to farmers, and from the Paleolithic to the Neolithic Age. The three emergences are not identical, but they may be related. Better tools may have made tilling the soil easier and hence aided agriculture. Social relations of settled communities may have provided the circumstances to develop new tools and new technology.

About 6,000 years ago changes began that transformed Neolithic societies into civilizations. This transition was remarkably rapid in some places. I first became aware of this when visiting the antiquities museum in Cairo. There are 30 dynasties covering 3,000 years or more. One can walk through these in reverse chronological order, noticing that each dynasty had its own culture and artifacts. Finally, one comes to the period before the First Dynasty, with its artifacts of Stone Age villages. The transition to the sophisticated First Dynasty can only be a couple of hundred years, but the consequences were enormous.

In Egypt and in cities of the Fertile Crescent, high population density led to centralized authority and stratified societies, and a new mode of social organization emerged. The changes of the urban revolution were as profound as the agricultural revolution. Large-scale field agriculture with an irrigation system was necessary to support the urban centers.

Socially the results of the urban revolution were profound. Armies and police units were organized. Religious institutions brought temples, a priestly class, and the accompanying higher education. Studies of medicine, engineering, and astronomy developed. Monuments and large engineering projects became hallmarks of urbanization.

The age of biological emergence is replaced by social emergence. Not only is there a new species, but a species that changes the environment to be dominated by artifacts rather than by natural ecosystems. The planet had been changed by the biosphere for four billion years, but the origin and spread of the technosphere was an entirely different order of magnitude. Urbanization only occupies about a one-millionth part of the age of the planet: our society is a very late arrival on the face of the Earth.

The Holocene is subdivided into the dominant technologies: the Stone Age (Neolithic), the Bronze Age, and the Iron Age. The Bronze Age (from roughly 6,000 to 3,000 years ago) began as the Copper Age, utilizing metallic copper deposits. The invention of bronze, an alloy of copper and tin, followed the use of copper. The Iron Age began about 3,000 years ago. The developing knowledge of metallurgy altered the conceptual definition of tools, making them much more sophisticated devices.

The devices of early technology were almost entirely mechanical. Somewhere around 1200 C.E. the discovery of explosive powder introduced a chemical aspect of tools. In the 1200s, the invention of the steam engine added another component to toolmaking. In the 1800s the understanding of electricity and magnetism resulted in electric motors and a whole new aspect of toolmaking. From here on the development of technology went explosive. The technology of building construction was the basis of urbanization, and the size of villages and cities grew from hundreds into tens of thousands.

Writing provided a technology of information storage. Formal pictographic writing probably goes back over 5,000 years, to be followed by syllabic writing in the Fertile Crescent. After the development of writing, there was constant intercultural exchange among the various societies. A fully alphabetic writing seems to have developed in Greece about 2,800 years ago, and the system has become almost universal. This is only 100 generations into the past.

Starting with movable type in the later 1400s, a new kind of machine appeared, one that basically dealt with information rather than with matter or energy. The development of information machines grew very rapidly in the 1900s, the age of computers. With very high-speed computers and robots and the domain of intelligent machines, some have suggested that a new kind of life is emerging. I think not, but will defer that discussion until we discuss mind and think about epistemology.

In summary, the earliest villages were semi-permanent, having to move when the surrounding soil became exhausted for agriculture. Ready access to fertile land was a requirement for each village or town. In the period some 6,000 or so years ago, agriculture and transportation improved to the point where cities were possible. As we noted, new kinds of activities—craftsmen, soldiers, priests, and others—became possible. The rise of domesticated animals for transport increased the possible area for a city. The wheel was first used in the Tigris-Euphrates Valley about 5,500 years ago. It opened up much more extensive transportation and created a need for roads. Cities began to develop all over the world, particularly on the shores of rivers.

It is clear that the rate of development of technology is constantly accelerating. After the Neolithic Age, the three millennia of the Bronze Age produced sophisticated tool kits as well as a class of artisans skilled at metallurgy. The next two millennia, the Iron Age, resulted in far more sophisticated tools. The scientific understanding of materials and energy for the next few hundred years set the stage for the explosive development from blacksmiths to computer programmers in less than 200 years. We are moving so fast that it is difficult to know where we are going. This is part of the novelty of emergence.

The role of technology in science is summed up in this statement: steam engines have taught us more about thermodynamics than thermodynamics has taught us about steam engines.


## Chapter 30—Readings

McClellan, James E., and Dorn, Harold, 1999, *Science and Technology in World History,* The Johns Hopkins University Press.

# 31

# Philosophy

The last series of emergences (toolmaking, language, agriculture, and technology) were related to survival in nature and competition with other hominids. The domains span the range from the biological to the cultural. Along the way something uniquely human was happening. As knowledge and understanding of the natural world expanded and deepened, much of it needed for urbanization, a nagging question began to rise in a variety of forms, all of which translate into "What does it all mean?"

The earliest science must have related to the regularities of nature: the dawn each morning, the succession of the seasons, the periodicity of the moon, the apparent movement of the planets and stars across the skies. Horticulture requires an understanding of seeds, the growth cycle of plants, and sometimes the role of roots and shoots. Domesticating animals requires an understanding of the behavioral modes of various species. Collecting and transmitting knowledge developed into a necessary function in agricultural villages. All of the above is speculative. We have to reason from knowledge of modern humans in a variety of cultures and the historical record, which covers only about 6,000 years.

The forces that controlled the world beyond the regularities gave rise to another set of postulates dealing with the existence of forces or beings who exerted control. This is the beginning of religion. Seeing events in terms of the natural and supernatural must go back into the hunter-gatherer days. This early dichotomy remains with us. Natural law provides the basis for understanding the unfolding of the world: that is science. There are also forces that seem to go beyond natural law that have a purpose related to humans, somehow controlling events for specific needs of human society. They are gods or God.

Both science and religion deal with unseen and abstract forces that control the world and individual persons. Science deals with the complete regularities of the laws. Religion deals both with laws and with violations of these regularities, which are designated miracles. The science of emergence also deals with novelties in the regularities.

Between science and religion lies philosophy in the formal sense. Philosophy is the attempt to understand both how knowledge is obtained and the nature of that knowledge, not only of the natural world, but the worlds of art, music, government, and all manner of activities of human beings. Thus, ethics and esthetics are included along with epistemology and ontology.

All of these activities and thoughts have probably emerged over the past 10,000 years or more as the ultimate emergence of social humanness. The interaction rules are incredibly complex, and the pruning rules are at the moment largely unknown. But the emergent feature is the human mind rooted in neurobiology and cellular biology and going back to the fundamentals of quantum mechanics and pruned by human/human interactions. Human thought is a very social construct, and new ideas are population-dependent because one needs not only a novel thought but also someone to discuss it with. Ideas with only one proponent disappear.

The last 100,000 years or more have witnessed the development of material culture from chipped stone tools to supercomputers. A wealth of physical remains exists. Archeology is rich with artifacts that allow us to reason about the developments in technology, particularly for the past 5,000 years. The great museum of Cairo referred to in the last chapter contains the physical remains of 30 dynasties, from 3000 B.C. to the Roman conquest. In the country surrounding the museum are remains of great archeological structures. We can chronicle the past 5,000 years quite well, but understanding emergent human thought is far more difficult, particularly thought before the development of writing.

Artifacts reflect technology, which speaks to the underlying science and mathematics of entities such as circles and right angles. But the great structures—pyramids, temples, cathedrals, and the art of gods and goddesses—speak to the ideological accompaniments to the emerging technology. A yearning for immortality seems to have accompanied the engineering marvels. This occurs in many societies. I have been to the pyramids of Egypt, to the great stone structures of Machu Pichu, to the statues of Easter Island, and to many of the cathedrals of Europe, and I never fail to be amazed at how technology has been put to work for religion.

One characteristic feature of human thought is common both to science and religion. In both domains one starts with the phenomenological world and tries to formulate explanations based on features or entities that are not part of the phenomenological world. One difference is that the unseen of science is thought to be totally lawful, while the unseen of traditional religion is volitional. A second difference is the mode of validation, which is experimental and public in science, and self-referential and private in contemporary religion.

In any case, 12 billion years of emergences have resulted in self-aware humans who seem driven to look backward and forward in asking who we are, where we have come from, and whither we are going. In short, we ask, "What does it all mean?" Philosophy provides a context for raising these questions.

Religion and science seem to have emerged independently in China, the Mediterranean region, and the Americas. It is difficult to know how much informational interaction there was among various parts of the world. But the commonalities in isolated parts of the world suggest an orderly scheme of emergence in human society.

For the series of emergences we have been viewing from *Homo habilis* onward, there has been a different character from the preceding biological changes. Language and writing have introduced information into societies, and this is a major novelty.

In religious thought, the monotheism of the Egyptians and Hebrews led to a more orderly theology than the Pantheon of independent gods. Monotheism provides a closer relation between science and religion. If the monotheistic God is identified with the laws of nature, science and religion come together; however, this leaves out the volitional aspect of God, an omission to which we must return.

Having come to the domain of philosophy impels us to ask what sources of knowledge allow us to investigate each emergence. Our approach has been deceptively linear, and with respect to time it is linear, but the kinds of knowledge that enter each emergence are more subtle. Thus we must turn to epistemology, the discipline that asks, "How do we know?"

This question was asked by Immanuel Kant in his *Critique of Pure Reason*. He reasoned that knowledge begins with experience, of which there are two components: (1) sensations brought to the mind from interaction with the external world, and (2) ideas brought internally to the mind that influence our sensations. He has designated these as *a posteriori* and *a priori*. Knowledge starts with the mind. Whether dealing with *a priori* or *a posteriori*, the mind is the primitive from which all knowledge starts.

The world out there or the thing in itself (*ding an sich* of Kant) is unknowable in any complete sense. (Note that our view has changed over the years from solid objects to molecules, to electrons and nuclei, to elementary particles, to quarks, to super strings.)

From experience the mind constructs its view of the world. This is the job of science. Philosopher of science Henry Margenau called atoms, electrons, probability distribution, and the like "constructs," entities that the mind constructs to understand the sensory world. From the constructs, we go on to try to understand the world in a hierarchy of emergences. Constructing constructs seems to be uniquely human.

In the search for ontology (the reality behind cognitive events), there are two views. The naïve realists believe that the constructs are real, and mind is an epiphenomenon. The idealists believe that mind is real and the constructs are epiphonomena.

In the emergence approach, we have operated in a circular fashion. Taking the scientific approach, we have utilized the *a posteriori* and the *a priori* to construct a world view that then has moved from its most reductionist entities through a series of hierarchical emergences to the mind, the site of the convergence of the *a priori* and the *a posteriori*. In other words, we have started with the mind, the epistemic *sine qua non* or necessity, and have built a universe of constructs that are then used in an effort to try to understand the mind. The mind is much more of a primitive than are atoms and quarks. For us, the mind and the universe are not separable. This is not to deny the thing in itself that would be there without the existence of humans: it is to assert that the universe that we know is some combination of the universe in itself and the mind in itself that are interactive. The result (that the mind interacts with the universe and tries to reconstruct the mind as an emergent property of the constructed universe) may be circular, but it is where our search for understanding takes us.

There seem to be critical levels in the hierarchy, such as atomicity, but the circle itself seems whole. This is not to argue for an anthropic universe: the thing in itself doesn't require us. It is to argue that the *knowable* universe is anthropic.

Another feature appears: while the mind seems like an individual concept, it is a social one, the collective minds of those seeking to understand the universe. Descartes's dictum, "I think, therefore I am," ignores that the first "I," the mind of the thinker, is the result of interaction with parents, teachers, mentors. When we talked of the emergence of primates and hominids, social structure became very important in shaping the mind of the "I." Language itself, one of the tools of constructing understanding, is clearly a social entity. Much of Descartes's "I" is already in place when he or we come to do philosophy.

Reduction and emergence are tools by which we try to relate the mind of the philosopher to the minds of the cognitive psychologist and the neurobiologist. It is a remarkable circle of understanding.

The view of philosophy discussed in this chapter represents one set of emergences, the Hellenic or Western tradition. Asia has developed other intellectual traditions. F. S. C. Northrop, in his remarkable 1946 book, *The Meeting of East and West,* points to the radical differences in the intellectual traditions that have developed in parts of the world that have been isolated during the emergence of philosophy. The different systems are now in contact, and a new synthesis is in process. Technology leads to a globalization of philosophy. This is countered by tendencies toward localization and provincialism, as we noted in our discussion of niches. Business is also undergoing globalization, but other forces are clearly at work. It is difficult to deal with current emergences unless we understand the selection rules that are operative. The search for selection rules is one of our major tasks.

## Chapter 31—Readings

Kant, Immanuel, *Critique of Pure Reason,* Translation 1929, St. Martin's Press.
Northrop, F. S. C., 1946, *The Meeting of East and West,* Ox Bow Press
    Translation, 1929, Reprint, 1979.

# 32

# The Spirit

To generalize our understanding of the past 10,000 years, we follow the suggestions from Teilhard de Chardin. Teilhard argues that the emergence of mind along the evolutionary pathway of the hominids was as globally significant as had been the origin of life. Our epistemological discussion certainly reinforces this point. And just as life has spread out and covered the planet, he reasoned that mind also was globalized; just as life gave rise to the biosphere as its planetary embodiment, mind gave rise to the "noosphere," the collective mental activity of *Homo sapiens.*

I once regarded noosphere as a rather poetic term of the French savant, but now I see the World Wide Web as a reification or instantiation of the noosphere and consider Teilhard as an even more prescient thinker. Human thought is collective. The idea of the solitary truth-seeker ignores the extent to which we are social animals with a long infancy and maturation into some degree of independence. So it is that technology, philosophy, and religion are all social activities. The "truth" of science depends on agreement among the practitioners: we now are all part of the noosphere, the meeting of East and West, as well as the biosphere.

In this chapter we focus on what to expect as the next emergence beyond the noosphere. It is the move from the mind to something more spiritual.

When I planned this chapter on emergence of the spirit I never realized how difficult it would be from a purely academic point of view. For the other emergences I know (if sometimes only dimly) what has emerged, while here I was trying to look to the future to the next emergence. This violates my epistemological imperatives. So for the moment, allow me to be a speculative futurist to think about what emergence we may be in the middle of.

Teilhard spoke of two emergences, the material and the spiritual, and he was roundly berated from a thermodynamic point of view. He was troubled by the two energies because he knew that somehow or other there must be a single energy operating in the world. He expressed his problem in the following words:

> The two energies—of mind and matter—spread respectively through the two layers of the world (the *within* and the *without*) have, taken as a whole, much the same demeanour. They are constantly associated and in some way pass into each other. But it seems impossible to establish a simple correspondence between their curves. On the one hand, only a minute fraction of 'physical' energy is used up in the highest exercise of spiritual energy; on the other, this minute fraction, once absorbed, results on the internal scale in the most extraordinary oscillations.

The problem he stated can be formalized in the following way: if the change in the Gibbs free energy of a system is split up in its component parts, the change may be written:

$$\Delta G = \Delta H - T\Delta S$$

where $G$ is the Gibbs free energy, $H$ *the enthalpy,* $T$ *the temperature, and* $S$ the entropy. Now, $S$ is the informational part of the thermodynamic state of the system. It is as if the Gibbs free energy part could be split up into a material and a mental component. Teilhard was closer to the solution than he knew.

Things improved from the Teilhard perspective when E. T. Jaynes, writing in 1957, showed that in statistical mechanics the entropy of a system was a measure of the information we would have if we knew which microstate the system was in, starting with a knowledge of the macrostate. This followed from Jaynes's reconciling information theory and statistical mechanics. There is in the Jaynes formulation a certain noetic property to information-based entropy that rationalizes Teilhard's two energy concepts. Teilhard had also intuited something semiquantitative between the two energies, as seen in the previous quote.

Using the Jaynes and Shannon (information) formalization, we can calculate that a binary decision involves an energy equivalent of:

$$E = kT\ln2$$
$$= 4.14 \times 10^{-21} \text{ joules}$$

T is the absolute temperature, and k is Boltzmann's constant. This is indeed a minute amount of energy.

In physics, the human mind has become more a part of the formalism. A probability distribution becomes an event by interaction with a human observer, and entropy deals with a human observer's ignorance of the microstate of the system. The operative free energy in biochemistry is the Gibbs free energy, which is minimized as the entropy of the universe is maximized. Changes in Gibbs free energy involved changes in the enthalpy or normal energy as measured with a Calorimeter and change in the entropy that deal with an observer's knowledge of the microstate of the system. The entropy term has a mental aspect. Teilhard sometimes speaks of this as spiritual energy, and other times as mental. At this point he could have called the measure noetic energy, rather than spiritual energy.

In any case, the two energies exist in Gibbsian thermodynamics, as well as in the Teilhardian perspective.

Teilhard later talks about the spirit and goes beyond thought *per se*. He refers to the spirit of the Earth and the hyperpersonal. When one addresses the quest for the spiritual, it is some aspect of existence that goes beyond the biological (the second great emergence), and beyond the mental (the third great emergence) into the domain of something more psychic, *"a formidable upsurge of unused power."*

I think that there is a feeling ranging from the theists to the existentialists that we have not fully evolved or have not worked our way to what we may become.

The rigorous Darwinians will ask about the next evolutionary stage when *Homo sapiens* will be followed by some other more fit hominids.

The social Lamarckians will ask how we can change our society to achieve the greatest good for the greatest number. If we are indeed the transcendence of the immanent God, this is our calling.

Two new futuristic views have developed in recent years. The first of these argues that carbon-based life will be the precursor of silicon-based life that, because of potentially superior intelligence, will ultimately take over, with humans either eliminated or in a secondary role. This involves a discrete molecular break in the form of life, but not necessarily a break in informatics and the handling of information. Of course, what love means to a computer is an unanswered question.

The second futuristic view is a world in which genetic engineering is used for us to become the race of hominids we want to be. Technologically this will be possible, but are we able to be wise enough to avoid unforeseen consequences? And will we know what we want the engineered human beings to be?

Teilhard has a more teleological view of a new and final state. At this point I neither understand nor follow him.

I assume that something new will emerge in human society, and it will present us with undreamed possibilities in science and the arts. This emergence requires our efforts and requires something spiritual that goes beyond the mind. There will be a new emergence, and we will play a part in what that emergence is. That is our destiny.

## Chapter 32—Readings

Moravec, Hans, 1999, *Robot, Mere Machine to Transcendent Mind,* Oxford University Press.
Teilhard de Chardin, Pierre, 1959, *The Phenomenon of Man,* Harper and Row.

# 33

# Analyzing Emergence

The 28 emergences represent a continuous series going from the reductionist core of particle physics to the most noetic aspect of human thought. The first seven emergences are studied by cosmology, high-energy physics, astrophysics, geophysics, and geochemistry. After each of them, the universe or some part thereof is radically changed. Something novel has happened. There is a surprise. The first is the greatest surprise of all; there is something rather than nothing. It is difficult in the extreme to relate it to other emergences.

The second emergence says, not only is there something, it is structured. In the most abstract sense, it didn't have to be that way. Two possibilities are (1) nothingness and (2) something absolutely uniform and structureless. The next emergence starts with large-scale structure, but small-scale simplicity confined to hydrogen and helium. Nucleosynthesis, which is the interaction rule in stellar dynamics, leads to an explosive emergence of complexification, for it fills the expanding universe with many different kinds of things. Gravity, nonuniformity, and particle physics all come together, and the nuclides emerge. The universe is now made of many kinds of entities not previously present. Complexity has increased.

The next emergence occurs when parts of the nonuniform universe cool enough for nuclei and electrons to come together. The governing rules are quantum mechanics, and the overwhelming pruning law is the Pauli exclusion principle. This leads to an even greater complexification. For a consequence of the Pauli principle is the emergence of chemical bonding. Instead of being limited by the 2,000 or so nuclides of the 90 or so elements, the world of molecular entities can grow almost without limit, depending on the number of atomic entities of different kinds found in any interactive portion of the universe. At this level, chemistry enters into the description of the evolving universe. Stardust consists of various-size particles coming from stellar explosions and collapses. The entities that are involved in making second-and third-generation stars and solar systems are held together by atomic and molecular forces. Emergent properties at the global and cosmic scale result in part from action at the atomic scale. This is even more true at the level of planets and other relatively low-temperature space debris.

In the formation of solar systems, the previous emergences come together. The sun, which as a second-or third-generation star still contains hydrogen and helium from the first emergence, condenses by gravity that powers the second emergence, operates by nucleosynthesis following the next emergence, and contains and is surrounded by the chemical elements that are themselves emergent. As the planets emerge from stardust and other cosmic debris, the rules are generated from celestial mechanics and geophysics and are dominated by fluid flow and thermodynamics, particularly the Gibbs phase rules. The shell structure of the newly formed planet seems to follow from the laws of chemistry and physics. Indeed, there is nothing in these steps to suggest anything beyond the laws of physics and chemistry, except to realize the enormous organizing potential inherent in the Pauli principle. It is no accident that we keep coming back to this principle. I believe it gives us a clue to the kinds of ideas to look for at other levels of emergence in understanding ideas that we don't yet have a good grasp of. The last of the astrophysical and geophysical emergences is the organization of the surface of the Earth into lithosphere, hydrosphere, and atmosphere. This is little more than a restatement that the surface must have a solid, liquid, and gaseous component; nevertheless, this feature is at the core of meteorology and climatology, which feed back on the development of the planetary surface. The geospheres have not been constant. The composition of the atmosphere has changed over 4.5 billion years from somewhat reducing to very oxidizing. The salinity of the oceans has changed, and the continents are in constant motion with all the tectonic consequences of this motion. Material cycles among the various shells, and the surface of the planet undergoes constant change. A complex planetary surface constantly changing in time has emerged.

The next emergence is the chemical organization of the planet—a kind of planetary metabolism leading to biological metabolism. Let us examine the features of the planet that drive the chemistry of interest. There are massive amounts of water with its unusual chemistry. There is carbon dioxide in the atmosphere and in the hydrosphere, and in the lithosphere carbon occurs as calcium carbonate. There are redox couples generated by the heat-driven cycling in the magma. There is an energy flow of solar photons. The planet radiates thermal energy to outer space. Thus, there are two major energy sources, sunlight and radioactive decay, both leading to the generation of oxidation-reduction couples. There are mineral surfaces such as pyrites that may allow for heterogeneous catalysis.

At this stage a set of organic molecules emerges. There is considerable difference of opinion among researchers in this field as to which molecules emerged and whether membrane-forming molecules preceded or followed metabolic and catalytic molecules. In one way or another, favored auto-catalytic chemical networks emerged, and the method of complexification changed to making information-rich polymers from a restricted set of monomers. The selection rules are ultimately properties of reaction networks, and the structure and catalytic features of macromolecules. Prokaryotic cells emerged. This emergence was the transition from chemistry to biology. For once in the domain of distinguishably self-replicating entities, competition takes over, the world becomes Darwinian and fitness becomes the pruning rule for emergence.

Emergences 1 through 7, from the primordium to the geospheres, are governed by the rules of physics treated in the broadest sense. The next emergence, metabolism, is intermediate between physical chemistry and biology. For subsequent emergences in biology, we want to go beyond fitness in a vague sense and try to look at deeper principles governing transitions.

The emergence of prokaryotes embodies a partitioning of cell interior and exterior by a barrier showing phase separation. This notion of an aqueous interior, nonpolar barrier, and aqueous exterior remains in force through all subsequent biological emergences. Metabolism, information storage in linear polymers, and the programmed synthesis of macromolecules also emerged with the prokaryotes. Mutation in the genome led to a new kind of complexification, a world of microorganisms with distinguishable features; speciation entered the unfolding of the Earth's history.

For the next emergence, the agents are species of prokaryotes, the rules are combinations of individuals of different species by endosymbiosis, and the pruning is by competition among all the varieties of chimeras. The new complexification is by combining those features that different species had succeeded in evolving individually. The emergence is of the eukaryotes.

Multicellularity is the next emergence. Since the cells are still the basic units, devices to stick cells together are required. From a biochemical point of view, the evolutionary novelty is the synthesis of sticky-surface molecules. There are presumably glycoproteins and perhaps lipoproteins; however, a uniform stickiness will presumably lead to a spherical clump of cells, which will deprive the interior cells of food and oxygen and will be of little advantage from a fitness viewpoint. *A successful multicellularity demands a morphogenesis.* This requires a language of types of sticky spots, rules of sticky-spot interaction, and differentiation of cells by function and by sticky spots to put the right function in the right place in the organism. Analogous to the genetic code, this is the morphogenetic code that also must be keyed to time and appearance of adhesion activity to give rise to developmental morphogenesis. This is studied in the molecular biology of morphogenesis, a new field of research. The previous sentences are somewhat speculative, but the uncertainty is partially demanded by the logic of what it takes to generate multicellularity. Fitness governs the selection.

Meaningful multicellularity requires a generation of cell types—thus, the emergence of specialized tissues and organs such as the liver and structures such as bones and shells. We have chosen the neuron as an example of the kind of emergence by cell type. It is, of course, a different type of complexification, but ultimately it leads to the millions of species of organisms. The neuron emergence is particularly important because it is on the main pathway to the ultimate emergence of brain, mind, and higher cognitive function. The general process leads to hundreds of cell types in evolutionarily later organisms.

The next series of emergences leading to the group of chordates, vertebrates, amphibians, reptiles, mammals, arboreal mammals, and primates is the classical Darwinian domain. It is here that emergence theory and evolution operate in the same arena. The agents are genes and gene clusters, but unlike the earlier agents, these get their meaning through an elaborate series of operations. Hence genomics, proteomics, and physiomics are all involved in the agents' acquiring their significance. The whole hierarchical problem puts a large gulf between genes and organisms. This makes emergence far more complicated, but doesn't change the fundamental nature. In part, the appearance of novelty lies in the ill-understood intermediate steps.

At this point we might take another look at emergence as viewed from computational science. In *Emergence,* John Holland has an extensive discussion of Arthur Samuels's work on checkers-playing programs. The program developed to play checkers must follow the rules, have a strategy for evaluating moves, and be capable of learning. The last of these involves changing weightings and strategies on the basis of experience. Experience consists of playing checkers with humans and other machines. A set of algorithms emerges, somewhat opaque to the program designer, but generally capable of defeating the original programmer in games of checkers. Fitness consists of winning at checkers. Again, note that the program defeats the designer by a set of complex intermediate computer operations unclear to the designer. Something truly novel has come about.

The relation between biology and checkers is that the intermediates between genes and organisms have something of the same characters as the intermediate strategies of checkers. The attempts to understand the intermediates from protein folding to physiological control is part of what goes on in many fields of modern biology.

The fitness of species is not a fixed quantity, and changes as geological, oceanographic, and climatic factors change the habitats. It also changes due to the presence or absence of other organisms. All evolution is coevolution, and the geospheres are ever-changing elements of an organism's habitat. Nevertheless, for the last billion years the Darwinian model holds at least to the arrival of higher primates, organisms that could think about their environment and then do something about it.

The importance of planetary conditions is shown in the emergence of the mammals as the dominant species. The triggering event was the great meteor hitting in the Yucatan. The cloud that covered the Earth resulted in a massive decline of the dinosaurs, and the resulting condition left niches for the mammals who had previously been minor species.

The series of emergences following the great apes has a different character. There is a shift to learned information between the generations and between individuals having a larger and larger role. The pace of change speeds up. Language, writing, printing, and computers each alter the character of emergence. The agents are now individuals, the rules are interaction between individuals and between individuals and the society, which at the beginning was a small group and now extends to the global population.

At the level of agriculture, overwhelming control of the environment by a single species emerges. The fitness rules are not just imposed by the habitat; the actions of agents, who are now groups of informationally interacting individuals, reacts back on the habitat, constantly altering the boundary conditions. With urbanization, affairs are moved up one hierarchical level and the state emerges, with all its institutions such as religions and universities.

Emergence has in an orderly way moved from protons to philosophers.

At this level there is a kind of closing of the loop, because philosophers think about Big Bangs, protons, and all the other hierarchies connected by emergences. The emerging world turns inward and thinks about itself. As George Wald once said, a physicist is the atom's way of thinking about atoms.

The communication of all the people of the world is perhaps what Teilhard de Chardin meant in his term, the noosphere. That very abstract notion is reified in the World Wide Web.

There is every reason to believe that there will be a next emergence, and I think that candidates are on the horizon, possibly in competition. The first concentrates on robotics, genetic engineering, and nanotechnology and looks to a world in which silicon life takes over from carbon life because information density and speed of processing give machines a distinct advantage. A whole literature is appearing on this emergence, such as Hans Moravec's *Robot*. This is being pursued by some of our most imaginative thinkers. To some it is a source of great concern.

The other view, introduced by Pierre Teilhard de Chardin 60 years ago, argues that the next emergence following the noosphere will be the emergence of the spirit. Trying to be true to both his passions, he tried to identify the world of the spirit with the world of his Jesuit brethren. The emergent world of the spirit need not relate to past theology, but may introduce novelty. Emergences are difficult to predict before they happen.

## Chapter 33—Readings

Stebbins, G. L., 1982, *Darwin to DNA: Molecules to Humanity,* W. H. Freeman and Co.

# 34

# Athens and Jerusalem

Thus far we have conducted an all-too-hasty review of the emergences from the Big Bang to our thinking about the Big Bang and what it means. We now look at the historical background that has given us tools to ask these questions and carry out our quest for meaning.

On a spring day in the year 350 B.C., there was a stir in the Academy at Athens. The increasingly independent experimentalist Aristotle was debating with the seminar mentor Plato, who adopted a much more theoretical view. Forty-nine years earlier, Socrates had downed his hemlock and thus ended his "misleading" the young of Athens. Two hundred miles away in the palace of Phillip of Macedon, six-year-old Alexander was beginning his education. Some 400 miles distant in another academy, this one in Jerusalem, a group of scholars was debating the meaning of a biblical passage. The Hebrew Bible had been codified and made canonical some years earlier, with the writings of Ezra and Nehemiah as the last included work.

The scholars at Athens and Jerusalem at that time knew little of their counterparts, although both were developing comprehensive world-views of God, man, and the universe, which would resonate far into the future. As Alexander matured and conquered half a world, including Jerusalem, the Greek and Hebrew views were finally to come into a dialogue that has lasted for well over two millennia and has a current importance that can be impressively seen in writings like the 1998 papal encyclical *Fides et Ratio* on the relationship between faith and reason.

It is our contention that the concept of emergence that has come out of complexity theory in just the last 30 years had much to say about dialog that has been proceeding since 300 years before the birth of Jesus. We will try to place it in context.

To view matters in the contemporary mindset, we turn to the article, "God, Concepts of" in the *Encyclopedia of Philosophy*. The first "problem" set forth is the immanence and transcendence of God. The immanence traces back to the "unmoved mover" of Athens, and the transcendence goes to the mysterious events on the mountain in the desert of Sinai.

In Abrahamic religions, God is transcendent, wholly other than the created world. The interactions between God and the world are through miracles, through special agents such as angels, and in Christianity through the more elaborate features of a trinitarian God. The essence of the transcendent God is unknowable. The stress on immanence found in Spinoza identifies God with the substance and laws of nature. Thus through the study of nature, some insights into God's essence are possible. This duality goes back to the God of history and God of reason, ideas that can be dated back to Alexandria at the beginning of the Common Era.

The God of Aristotle was the "unmoved mover" powering the world and contemplating his eternal verities. This concept matured into St. Thomas's God of reason 1,500 years later. The God of Jerusalem was Jahweh, the brooding spirit over Sinai, who guided his people to Canaan and raged against their unfaithfulness through the voices of prophets. In Christianity, this God changed into the God of faith of St. Thomas Aquinas. But in Judea and Israel, he was better known as the God of history. The God of Sinai speaking to Moses "mouth to mouth" went beyond faith. This God was a historical presence. The faith was involved in believing those who reported the words of the Lord. For Christianity, it was belief in those who reported the events of the resurrection, and for Islam it was belief in the Prophet's reports of the word of the Angel Gabriel. Faith is a more epistemologically abstract concept than was of concern to those who followed the views of Nehemiah.

Following the vision of Paul on the road to Damascus, the God of History became the God of Faith for the developing Christian church. Paul had no direct contact with the historical Jesus, therefore his conversion was totally an act of faith. Much of subsequent Christian philosophy is an attempt to reconcile the God of Faith with both the historical God of the Jews and the philosophical God of the Hellenists. The trinitarianism of Christianity also embodies three views of God (Father, Son, and Spirit); although there is not a direct parallelism between the two triads, there are common features. Indeed, 2,000 years of theological disputation leaves the matter still unresolved.

In 350 B.C. the God of Athens and the God of Jerusalem had little to do with each other, although Jews and Greeks may have encountered each other in Babylon and Persia and other sites. Twenty years later, the armies of Alexander burst out of Macedonia and carried Hellenic culture to North Africa, to the Judean countryside and into the Balkans. Alexander brought the God of Athens to the gates of Jerusalem and beyond and established at the mouth of the Nile the city of Alexandria, which became the "Athens" of the pre-Christian world. It shortly became home for a large number of Jews and the international site of major interaction between Hebrew and Greek culture. The Jews of Alexandria spoke a dialect of Greek known as Septuagint Greek.

During the reign of Philadelphus, the second Ptolemy to serve in the western district after the breakup of Alexander's empire (285–247), the Pentateuch and other parts of the Hebrew Bible were translated into the Greek of the Alexandrian Jews. By the time of Philo, the entire Old Testament was available in Greek. The theologies and philosophies of Athens and Jerusalem were now in an interaction that has affected all of subsequent Western culture. The Old Testament in the post-apostolic age probably entered Christianity primarily through the Septuagint version.

A century after the translation, we find fragments of the Greek Bible in the writings of Aristobulus of Paneas discussing the "book" in terms of ideas from Pythagoras, Plato, and Aristotle. Methods were developed to reconcile the anthropomorphism of the Pentateuch with the philosophical concepts of God. Paneas was in northern Palestine. Aristobulus is the first author I have found who discusses the philosophy of Athens with the theology of Jerusalem. That dialog has resonated for the next two millennia and gives no sign of letting up. I believe that the concept of emergence will cast some light on the issues under discussion by Aristobulus.

In Judea itself, the triumph of the Maccabees led to a suppression of Hellenistic thought. And although some Greek ideas crept into the Talmudic writings of the Rabbis of Palestine and Babylonia, the major action in that sphere was in Alexandria, where a Greek-speaking Jewish community was in constant contact with the Hellenic culture planted on the shores of Africa by Alexander. Philo tried to blend abstract ideas of God from Plato with the more tangible historical God of the Pentateuch. He used an allegorical interpretation of Scripture to remove inconsistencies between the biblical God and Greek thought.

Philo accepted God's creation of the laws of nature, but allowed the historical God to infringe on his own laws for the benefit of mankind or individual humans. Philo also allowed for free will, man's powers to direct the laws of his own nature. Philo appears to have a dual epistemology, with knowledge from sensation and rational explanation and knowledge of a prophetic character directly from God. These permitted us to know God's existence, but not his essence, because it was so unlike anything else in the world.

Philo's words were welcomed by the early church fathers, for whom the synthesis of Hellenistic and Hebraic thought was the core of the emergent Christianity. His influence is seen in particular in Clement of Alexandria, Origen, and Ambrose.

In Philo we have the synthesis of the previous centuries of Western philosophical and religious thought. The synthesis was of course incomplete, for the argument still goes on. I think that many scholars and historians have not appreciated the significance of the two Gods of Alexandria and the God of Paul in formulating the Western view of the world. This brief history of thought we are engaging in is to set the intellectual milieu in which emergence may offer some insights.

One hundred years after Philo, Titus Flavius Clemens (Clement of Alexandria, 150–213), an early Christian theologian, followed the thoughts of Philo. As one of the church fathers, he added Christian truth (the word of the New Testament) to the theology of Philo.

There is a difference between Judaism that accepts the historical truth of the Old Testament by unbroken cultural tradition, a passing on from Moses to the present, and Christianity, which starts from acts of faith, "acceptance from God of an undemonstrable first principle from which all other truth may be deduced." With Clement a view arose that was neither Athens nor Jerusalem, but can be traced to the road to Damascus.

Clement was preceded by St. Justin, whose martyrdom in 165 in Rome stems from his teachings of Greek philosophy and Christian theology. Clement was followed by Origen, who developed Christian systematic theology.

Saint Gregory of Nazianzus produced a literature to clarify the nature of the three parts of the Trinity. This problem of Father, Holy Spirit, and Son existed since the beginnings of Christianity, which arose from the passionate monotheism of Judaism and required a special role for Jesus that went beyond prophecy and political messiahship. The nature of the Trinity was at the root of the Arian heresy in the time of Constantine, festered for six centuries and exploded in the breakup of the Christianity of Rome and Constantinople in the eleventh century. It remains a source of contention among present-day Christian communities and resonates in battles in the Balkans.

For the pre-Hellenistic Jews, the anthropomorphism of the Bible was adequate, and no bridges were required between God and man, other than the angels and prophets. By the time of Philo, a contemporary of Jesus, the direct link was philosophically unsatisfying and a pathway was required for God to intervene in the affairs of man. The more the Hellenistic philosophy demanded a less anthropomorphic God, the more serious the divide. This was not only a Christian problem, as Philo, Sadia Gaon, and Maimonides faced it in Judaism, and various Islamic philosophers dealt with the issue. The problem has been to personalize God for man. The bigger and older the universe became, the more difficult this became. Monotheism and trinitarianism have to face the route of contact between man and God. It is difficult to be warm and personal about the God of the Big Bang. We will later return to this issue.

These matters were faced by St. Augustine, who was able to solve philosophical problems by a prior acceptance of scriptural truth as the core of his epistemology. This provided a background for the next 800 years of church philosophy.

This all too brief précis of classical religious thought is to remind us of the roots of the theological ideas that must be considered if we are to open a dialog between religion and science. Underlying the conversation is knowledge by reason, knowledge by historical tradition, and knowledge by faith. To these, the scientists would add knowledge by empirical verification and falsification. A new element, knowledge by modeling the world, must now be added to the previous sources.

In the seventh century a novelty entered the world of thought, the rise of Islam as a major world religion. In the ninth century, Abu-Yusuf al Kindi and his colleagues in Basra and Baghdad, following other Arabian scholars, continued the translation of Aristotelian-Neoplatonic works from Greek to Arabic. This also introduced knowledge by reason, which was considered inferior to knowledge acquired by Scripture and prophecy.

There followed a golden age of Arabian philosophy, which began its decline with the critical works of al Ghazali (1053–1111), whose books (such as *Incoherence of the Philosophers*) were intended to minimize knowledge by reason as compared to knowledge by faith and history. Nonetheless, the issues of faith, reason, and history were common to all the Abrahamic religions.

The last century of this age of Islamic philosophy was centered in Spain and marked by the works of ibn-Rushd (Averröes), which appear to have been of much greater influence in the Christian West than in the Islamic world.

The twelfth century was the time of Maimonides, a Jewish philosopher living in the Moslem world. He was thoroughly acquainted with the works of Aristotle and was in the tradition of Philo in preserving the truth of Scripture when in conflict with reason by assuming the allegorical interpretation of scripture. In trying to be a supreme rationalist, he moved in the direction later epitomized in Spinoza in looking for a rational interpretation of miracles. When he neared the edge, he always returned to Scripture. He reasoned that Scripture contained the essence of all metaphysical truth, enveloped in more or less allegorical wrapping. However, there are Orthodox Jewish groups today who consider that he passed the line into heresy.

The medieval period can be characterized as the struggle between faith and reason and attempts to reconcile the two. The best-known scholar of this surprisingly interactive group of Christians, Jews, and Moslems is St. Thomas Aquinas. He fully integrated Aristotelian philosophy into the mainstream of Catholic philosophy, subject to the priority of faith. This is a major problem in all ecumenical efforts.

The problem with the continuing return to faith is that Scriptures always have a historical component, and every faith has a different history, so that a universal faith is precluded by the very processes that generate the beliefs.

The multiple heresies of this book consist of rejecting history as a source of faith, and instead using history to study emergence, which can generate beliefs. Thus we start with divine immanence, which is almost a universal belief. We then study emergence as a domain of science and philosophy. We then argue that emergence has a divine aspect: it is the process by which the word (immanence) becomes flesh (transcendence). Emergence has a historical aspect, but may be more universal than the historical events recounted in Scripture. Thus there is some hope that the transcendence will be transnational and transdenominational.

As this book has proceeded, I have become impressed by the extent to which ideas about emergence have taken us back to the Scholastics, that group of scholars from 800 to 1500 who invested enormous effort into trying to rationalize the God of history thundering over Sinai, the God of reason of the Academy at Athens, and the God of faith appearing to Paul on the road to Damascus and to Mohammed at Medina.

In retrospect, the extraordinary feature of this period was the extent to which Muslim, Christian, and Jewish philosophers were asking the same questions, and the extent to which they listened to each other's voices. The

TABLE 8: MAJOR THINKERS OF THE HISTORICAL AGE

| | |
|---|---|
| Aristotle | 380-328 B.C. |
| Codification of the Bible | ~350 B.C. |
| Alexander the Great | 356-322 B.C. |
| Septuagint translation of the Bible at Alexandria | 250 B.C. |
| Maccabean Revolt | 168-164 B.C. |
| Aristobulus of Paneas | 160-? B.C. |
| Philo of Alexandria | 15 B.C.-60 C.E. |
| Saint Justin | 100-165 C.E. |
| Origen | 185-254 C.E. |
| Saint Gregory of Nazianzus | 330-389 C.E. |
| Saint Augustine | 354-430 C.E. |
| al Kindi | 850 C.E. |
| al Ghazali | 1058-1111 C.E. |
| Averröes | 1126-1198 C.E. |
| Maimonides | 1135-1204 C.E. |
| Saint Thomas Aquinas | 1225-1274 C.E. |
| Galileo Galilei | 1564-1642 C.E. |
| Baruch (Benedict) Spinoza | 1632-1677 C.E. |
| Isaac Newton | 1643-1727 C.E. |
| Charles Darwin | 1809-1882 C.E. |

series of translations between Greek, Arabic, Latin, and Hebrew speaks to this common interest.

With the beginning of science, Scholasticism was coming to an end. If Galileo was at the beginning of science, Spinoza is a related end to Scholasticism.

Since the ideas we discuss cover over 2,000 years, dating certain major thinkers in Table 8 provides a historical context.

In a sense, the spirit of this book is somewhat medieval, for that was the last period when science and theology were viewed as part of a holistic view. In the Renaissance, the two approaches parted company and either argued or ignored each other up until quite recently. I suggest that we are now in an era that is ready for a dialog between faith and reason. I, for one, welcome this return to the Scholastics, although I remain a card-carrying scientist. The dialog is important.

## Chapter 34—Readings

Bentiwich, Norman, 1910, *Philo-Judaeus of Alexandria,* The Jewish Publication Society of America.

# 35

# Science and Religion

The dialog between science and religion has often been extremely simplistic. The epistemology of faith has led some theologians to a totally literal interpretation of Scripture, which opponents attempt to refute when it disagrees with the scientific view on issues such as age of the Earth. For some scientists, the fear of teleology had led to a cavalier assumption that each step along the history of life or even before represents an accident, so that the whole unfolding in time is a matter of chance. I contend that both views are wrong and make dialog virtually impossible.

I hope that the previous chapters have demonstrated that the emergences are not completely matters of chance, but are governed by physics, chemistry, geophysics, ecological principle, and other laws of science that reduce the universe of chance to zones of the probable. We are far from a complete understanding of the pruning rules, but we know that they operate at every one of a large number of hierarchies, and they do not violate the underlying laws of the physical science.

The first break between science and religion covers the period shown in Table 9 below.

Copernican astronomy placed the sun at the center of the universe in disagreement with the biblical view, which had the Earth at the center. The next 150 years culminating with Newton's *Principia* witnessed the development of the concept of natural law that governed the motion of bodies both on Earth and throughout the celestial system. The church tried to suppress these ideas through the execution of Bruno and the house imprisonment of Galileo.

With Spinoza's essay "On Miracles," the argument was fully engaged.

TABLE 9: SCIENTISTS IN THE
CONTROVERSY WITH
RELIGION

| | |
|---|---|
| Nicolaus Copernicus | 1473-1543 |
| Giordano Bruno | 1548-1600 |
| Johannes Kepler | 1571-1630 |
| Galileo Galilei | 1564-1642 |
| René Descartes | 1596-1650 |
| Benedict Spinoza | 1632-1677 |

Spinoza identified God with the laws of nature and then argued that miracles are violations of the laws of nature. This left God violating his own laws, which Spinoza found impossible. Philosophers from Philo to St. Thomas had worried about the same sort of thing, but the laws of nature by the time of Spinoza were better formulated, causing a more severe problem. If God operated through a set of inviolate and deterministic laws, then God was removed from the world of mankind.

In part, the success of Christianity was the role of Jesus as an intermediary between the stern God the Father and mankind. If God the Father is a set of unchangeable laws of nature, the existence of the intermediary did not solve the problem of the isolation. Both the scientists and the theologians were quite comfortable with the idea of the creator, but this does not solve the problem of the day-to-day relationship between man and God. The creator, the mysterious immanent God of the laws of nature, is too far removed from our everyday lives. As the universe got larger and older, the immanent God became more and more distant.

Our review of the emergences has shown us that the unfolding of the universe is not totally determined; neither is it totally random. The truth must lie somewhere in between. We have to give up simplistic approaches. The world is far more complicated than was envisioned by earlier philosophers.

The status of the selection rules is not well understood. The Pauli exclusion principle looks like a law of the immanent God, yet it is at the root of an emergence. The emergence of apes from Old World monkeys looks like a response to geophysical and meteorological factors changing large-scale habitats. There are many selection rules, at least one per emergence. We have not fully classified them or explained them. However, they are determinative in the unfolding of the world. Our task as scientists is to try to understand these rules and develop an epistemological understanding.

To the theological, the selection rules are at least the intermediate between God's immanence and the development of our world. The trinitarians would designate this as the Holy Spirit. It is the path by which the word (the laws of nature) becomes flesh. While there has been 2,000 years of argument about the relation between God the Father and the Son, there has been remarkably little attempt to understand the Spirit as intermediate between physical law and humanness. I argue that the understanding of emergence is not only vital to understanding science, it is crucial to a natural theology in the ongoing effort to seek the relation of the created to the creator.

Thus far we have been dealing with 15 billion years of emergence. Sometime over the last 5 million years, something radically different occurred: the emergence of a species capable of attempting to understand cosmic history and purpose and capable of altering some small portion of the universe in ways far more radical than anything in the past. We can only trace this history in a vague way, but it appears that, after two or more million years, the early hominids of the genus *Australopithecus* gave rise to *Paranthropus,* which went extinct about 2 million years ago, and *Homo,* one species of which still persists. About 2 million years ago also saw the appearance of *Homo habilis,* a toolmaking hominid who left evidence of his work. There is then a more or less continuous series leading to *Homo sapiens* about 200,000 years ago. The uncertainty in this record does not affect our main argument of a continuous development of brain size, manual dexterity, and social organization.

Agriculture, language, technology, war, and religion were major transitions. The last of these was part of an attempt to understand the world and to control it. Control was also exercised in the transition from hunter-gatherers to agrarians, with major alteration of forests and savannahs.

Twelve billion years of emergence finally led to a creature who had the ability and chose to ask, "What does it all mean?" Eating at the tree of knowledge seems like an inevitable consequence of the development of the universe. There is little doubt from current understanding that there must be a large number of planets upon which intelligent beings may be asking for the meaning of the universe.

In any case the laws of nature (the immanent God) operating under the rules of selection (the Spirit) gave rise to *Homo sapiens* and human society. In this context, the metaphor of man being made in God's image seems appropriate. The interaction of God and man still seems remote.

To move ahead, consider the two aspects of God, immanence and transcendence. Immanence is natural law, eternal, unchanging, remote from mankind. A transcendent God is outside of nature and natural law, yet responsive to the needs of humanity and capable of contravening natural law for the benefit of individuals or peoples. The transcendent God is very anthropomorphic, hearing prayers and answering them. The two views of God are logically inconsistent. We return to Spinoza's essay on miracles and think of a transcendent God violating laws he created as an immanent God. It is a paradox.

Note that God's transcendence was not meaningful before the emergence of humans and human culture. Violation of natural law is only meaningful to individuals capable of knowing natural law. Divine transcendence arose from immanence and emergence and coevolved with *Homo sapiens*. Transcendence is an emergent property of God's immanence and rules of emergence. We *Homo sapiens* are the mode of action of divine transcendence. Consider an example: an ill child is close to death with an infectious bacterial disease. In the classical mode, one would pray to the transcendent God to interfere with the disease process and cure the child. In the modern mode one would give the appropriate antibiotic to inhibit or stop the growth of the disease-causing bacteria. In both cases, there is a miracle. The natural process of bacterial growth is stopped in a specific way, and a life is saved. In the first case the transcendence interfered with the immanent process of bacterial growth. In the second, the transcendence is the power of the human mind to study and understand the process of bacterial growth and to devise nontoxic methods of interfering with that growth. Transcendence in this context means that, with the evolution of the human mind, we can generate new emergences that were not part of the presapient world of immanences and emergences.

The antibiotic example is a rather poignant one, and there are no limits to the process. Transcendence leads to agriculture to prevent starvation and to aerodynamics that permit us to fly. It leads to governments to allow us to live in peace with each other and electric lights that allow us to function at night.

But the kind of transcendence that comes with the human mind is a two-edged sword. The same kind of activity that leads to antibiotics can lead to germ warfare. With transcendence comes the awesome power to choose good or evil.

Choice emerges with consciousness. We have argued that the fitness of consciousness is that, given the huge variety of environments, one can distinguish far more states than can be encoded for. Making the fit choice then becomes advantageous. This is the beginning of free will. When it is finally combined with the ability to understand the consequences of interactions, our collective behavior becomes transcendence.

I am aware that this is a startling, frightening, and thoroughly heretical conclusion. If our evolving minds are the transcendence of the immanent God, then the responsibility of making a better world is ours, as is the responsibility of figuring out what we mean by a better world. Our exemplars, Moses, Buddha, Jesus, Mohammed, and many more are those who have struggled the most in the search for the path of life. We have no one to turn to except ourselves and our exemplars. We are the third branch of the trinity. We dare not turn away from the task. There are no limits. Computers and genetic engineering give us whole new pathways in our transcendence. Emergence is not through with us or our universe. We must celebrate our divinity and go on with the nitty-gritty of the world.

The views we are developing are sufficiently radical that they are worth repeating to avoid confusion. We start with the faith that the world has meaning. This leads us, following Bruno and Spinoza, to pantheism, "the doctrine identifying the Deity with the various forces and workings of nature." It is not possible to understand the evolution of the complex system of nature from the reductionist first principles. A second feature enters the rules of emergence, which connects the most reduced principle of nature with the actual world that we see around us. Among the emergences was mind, with the possibility of understanding the universe and the ability to respond in a nonprogrammed way to the multiple emergences that led to some degree of volition and free will. Once this happens, we are partners of the immanent God in directing the further unfolding of local events in time. The rules of emergence are presumably features of the immanence, but when volition sets in, something has changed; consciousness carries with it transcendence. We can change the world for the benefit of mankind. We, *Homo sapiens,* are the transcendence of the immanent God.

"We are God," the best and worst of us. The statement embodies such hubris that it is hard even to announce, but I believe it contains a profound truth. The immanent God is knowable to us through our science, and the transcendent God is knowable to us through our actions. It is not the God of our ancient and revered faiths, but the world has changed, and we too must change our thinking. The intermediate emergent, God, must be understood next.

## Chapter 35—Readings

Robson, John M., Editor, 1986, *Origin and Evolution of the Universe: Evidence for Design?,* McGill-Queens University Press.

# 36

# The Task Ahead

This book has proceeded with two agendas: to study emergence by examining a number of examples, and to seek for the nature and operation of God in the emergent universe. We thus have reviewed in a very general way a series of major novelties from the Big Bang to the Spirit. At each stage we have sought for underlying interactions, laws, and ways in which actual outcomes have been selected to form the complex world of the possible. The laws of physics and chemistry have been identified with the immanent God, a very impersonal God committed to a lawful universe. This is the nature of the God posited by Spinoza, Bruno, and Einstein. The immanence is unknowable except through a study of the laws of nature. We study God's immanence through science. I am sure that there are scientists and theologians who are uncomfortable with that statement but its truth seems undeniable.

Lawrence Henderson, in his book *The Fitness of the Environment*, argued that deep within the laws of physics and chemistry the universe is fit for life. This fitness we identify with God's immanence. To many, it seems like an emotionally unsatisfying view of the divinity, but it does permit us to know God by studying nature. The present study of this fitness takes place under the rubric of "design."

In the unfolding of the universe, each emergence selects the restricted world of the real from the super-immense universe of the possible. There are two general approaches. The theologically minded world says that these choices are design, the way that the hand of the Creator enters into the evolving world. Trinitarians would call this aspect of God the Holy Spirit, whereas strict monotheists would simply regard it as another aspect of the immanent omnipotent God. Atheists would maintain that the choices are random and then become reified as frozen accidents.

All of these groups discover natural happenings that are not totally explicable. The theologically oriented would point to these events as God's design within nature, but not deterministic and therefore involving divine interference with law. The atheists explain away the emergences as frozen accidents. These are quite opposite explanations that deny a third range of possibility. There may be a new, as yet undeveloped, science of emergences that allows us to understand the selection of the subspace of the actual and, within limits, to predict the behavior within these constraints. The emergences would then be more deterministic and closer to God's immanence.

As we have previously noted, the God of Western religions is volitional rather than totally deterministic. There are two ways in which determinism can fail: randomness inherent in the immanent laws of nature, or a God who violates the laws of nature to fit some divine purpose, presumably for some ultimate human end. We are just beginning our understanding of emergence, and hence must be patient about understanding how the Word becomes flesh.

The third aspect of the divine for Trinitarians is the Son, involving incarnation and resurrection. This is far more abstract than the demands of most Christian dogmas. For traditional monotheists, this incarnation is man being made in the image of God. For philosophers, this is God's transcendence, the volitional aspect of the divine, reified in the activities of humans.

For science, the emergence of volition can be identified with the emergence of mind. This does not have to be identified with the evolution of *Homo sapiens* in a single step; it may have begun earlier and may continue into some other form. Wherever we come upon behavior that is not strictly deterministic, it is random or volitional. We associate the minds-of-animals theme with the ideas developed in writings of Donald Griffin. As we move from earlier forms to primates to great apes to hominids to humans, this volitional aspect seems to become more and more prominent. In humans we might identify it with free will, which becomes an emergent property of primate evolution and perhaps of the evolution of other taxa.

Putting this in more theological and more shocking terms, the volitional mind of man is the transcendent emergence of an immanent God. We are made in God's image because we are totally constrained by the laws of nature (divine law). If we have free will, we are transcendent and can perform miracles, volitional acts that are not totally determined, even though they do not violate the laws of nature. This is possible because of complexity and emergence.

There are two classes of nondeterministic happenings in nature, random and volitional. Consider these examples:

1. A cosmic ray from a distant star passes through the maturing ovum of an elephant. This leads to ionizations that cause a mutation in the genome. This egg is fertilized, resulting in an elephant that grows up and 20 years later suffers from a lethal disease as the result of the cosmic ray.
2. Heavy rainfall leads to a rising river in a town. While most residents await the flood, one individual, in a display of leadership, persuades everyone to build a sandbag levee to keep out the floodwaters. The level holds, and the town is spared the great damage that would have ensued without the levee. There is no violation of the laws of meteorology or hydrodynamics, but the volitional act of one individual and its social consequences change the course of the waters. This is a miracle of transcendence. It is the kind of miracle potentially always available to us as humans.

Not all acts of transcendence are beneficent. The decision to destroy a town and murder its inhabitants is also a volitional act. The emergence of mind has the potential for good and evil. Emergences of spirit have the potential of optimizing the good in the world of humans.

If we and our minds are the emergent transcendence of the Immanent God brought about by the rules of emergence, and if we possess any measure of volition or free will, then the burden of optimizing the good and minimizing evil is ours and ours alone. We cannot yet stay the waters of the floods, but we can minimize damage. We cannot stop disease and aging, but we can work to cure illness and to ease pain and suffering. When we fail to cure, we can assure palliative treatment and compassion for the terminally ill. I think that we can sense what the emergence of the Spirit can bring. This is, of course, our utopian vision.

The immanent God is universal. The God of emergence may be less universal, since conditions vary from place to place in the universe and the emergent life is different. The transcendent God is cosmically local, depending on the historical series of emergences that have led to mind and volition. These will clearly have a local character.

We began by thinking about a dialog between science and religion, and we find a surprising number of points of intersection. We have not discussed the points of departure where religions require direct interactions between man and a very anthropomorphic God, or representatives of that anthropomorphic God. Transcendence for traditional religions is within God rather than being in us as emergent from God.

Judaism does depend on the historical Moses ascending the mountain and talking mouth-to-mouth with the Almighty. Christianity is based on the incarnation and return to the Divine of Jesus, who is "God" or "of God." Islam depends upon the angel, a direct emissary between the transcendent God and the prophet. Scholars going back at least to Philo have been uncomfortable and have sought metaphorical interpretations of sacred texts, but the mainstream appeal has been to a personal God who hears our prayers. This is where the dialog begins. I suspect that it will remain a dialog for a long time, but it is too important to let it go on. It is the point of difference between our religion of emergence and traditional religions. It is also the point of difference between various religions that come out of different historical roots.

To those who believe that we are the mind, the volition, and the transcendence of the immanent God, our task is huge. We must create and live an ethics that optimizes human life and moves to the spiritual. To do this, we must use our science, our knowledge of the mind of the immanent God. I am reminded of the words of the Talmudist: "It is not up to you to finish the task: neither are you free to cease from trying."

# Index